Praise for

DARK BANQUET

"I was totally absorbed by this thoroughly charming and scientifically accurate account."

—E. O. Wilson

"What starts out as a horror movie of a book morphs into an entrancing exploration of the living world. . . . You'll be spoiling one dinner party after another retelling Schutt's tales of bats, leeches, and bed bugs."

—Carl Zimmer, author of *Parasite Rex*

"The definitive account of blood feeding in nature."

—Michael Novacek, provost and curator, American Museum of Natural History

"Dr. Schutt's voyage through the world of blood feeders is alive with humor and the sheer fun of scientific exploration."

—Charles Pellegrino

"Schutt is an engaging writer."

—*Washington Post Book World*

"Wide-ranging . . . packed with fascinating facts and anecdotes."

—*Newark Star-Ledger*

"Schutt may have a doctorate in zoology, but he has also mastered the pen along the way. And, he gets an A+ for his droll sense of humor."

—*Roanoke Times*

"Fascinating . . . [Schutt's] conversational style and gentle sense of humor serve the topic well, allowing the painless delivery of some impressive lessons in biology, evolution, and how science works."

—*Bats*

"Schutt hooks the reader on the first page . . . riveting . . . well rounded . . . engrossing science."

—*Library Journal*

DARK BANQUET

Blood and the Curious Lives of Blood-Feeding Creatures

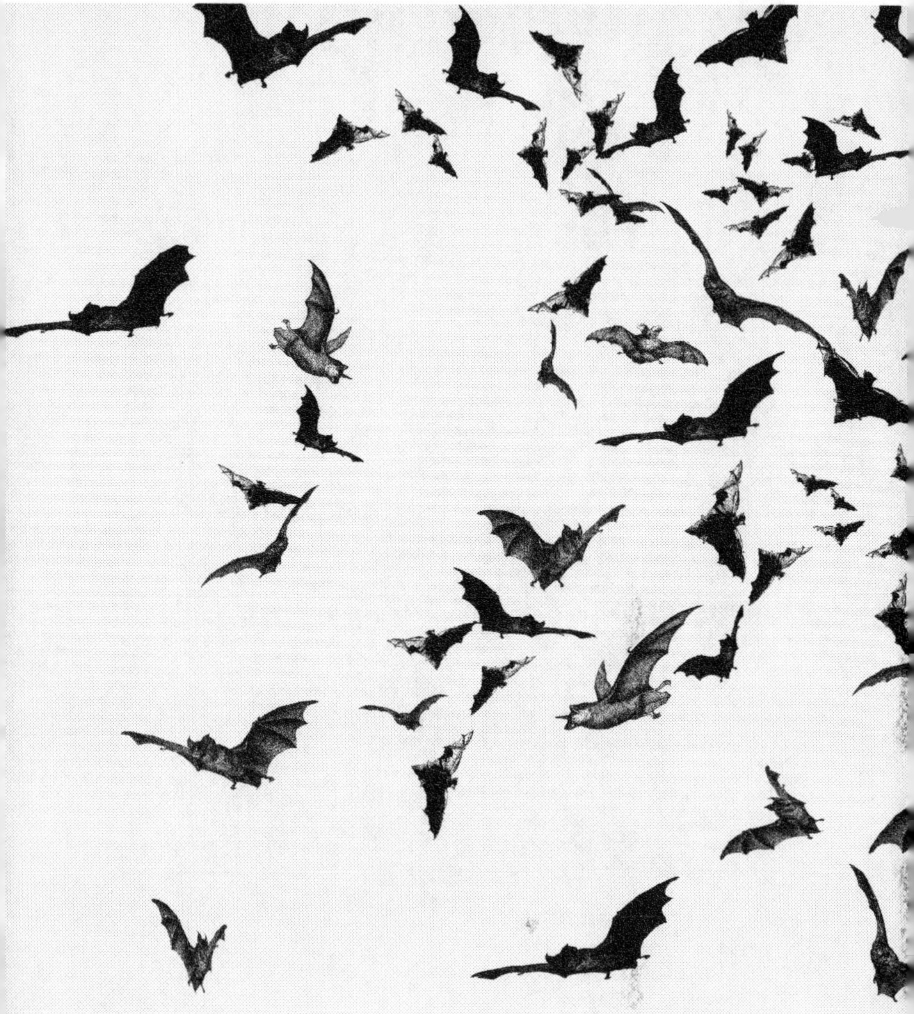

Bill Schutt

Illustrated by Patricia J. Wynne

THREE RIVERS PRESS
NEW YORK

Published in the United States by Three Rivers Press, an imprint of the Crown Publishing Group, a division of Random House, Inc., New York.
www.crownpublishing.com

Originally published in hardcover in the United States by Harmony Books, an imprint of the Crown Publishing Group, a division of Random House, Inc., New York, in 2008.

Library of Congress Cataloging-in-Publication Data
Schutt, Bill.
Dark banquet : blood and the curious lives of blood-feeding creatures / Bill Schutt; illustrated by Patricia J. Wynne.
p. cm.
Includes bibliographical references.
1. Blood-feeding creatures. I. Title.
QL756.55.S38 2008
591.5'3—dc22 2008003061

ISBN 978-0-307-38113-2

Design by Maria Elias

First Paperback Edition

146119709

For Marie Grace Schutt and William A. Schutt Sr.
. . . and all my Aunt Roses

CONTENTS

I know that our late King, though not apt to believe more than his neighbours, had no doubt of the existence of vampires and their banquets on the dead.

—Horace Walpole, commenting in a letter on the beliefs of King George II

The blood is the life.

—Deuteronomy 12:23

PROLOGUE

(2002)

A pair of chickens scratched nervously at the dusty ground beneath the grapefruit tree, careful to avoid the small puddles of coagulated blood.

"This happened last night." The voice from behind me belonged to Amos "Jumbo" Johnson, my guide and field assistant. Jumbo worked for Trinidad's Ministry of Agriculture in the Anti-Rabies Unit. I'd figured out several years earlier that Jumbo had gotten his nickname from the fact that the only thing he liked more than eating food was talking about it. But now he had gotten sidetracked—sort of.

"Tonight when the blood is fresh, it will glisten."

I nodded, trying to determine if either of these sad-looking birds had been bled the night before. *Was that a dark stain along one of their legs?*

It was my third trip to Trinidad and I'd come for the same reason each time: to study vampire bats, arguably the most highly specialized of all living mammals. Feeding solely on blood, vampires make up a tiny fraction of the order Chiroptera (only three out of the eleven hundred bat species). But even among this exclusive group, *Diaemus youngi,* the white-winged vampire bat is special. Far more rare than *Desmodus rotundus,* the aptly named common vampire bat, *Diaemus* is an arboreal hunter—feeding primarily on birds and currently subsisting almost exclusively on the blood of domestic poultry. This in itself wasn't all that strange. It was, after all, the arrival of man and his cattle that had exploded the common vampire bat populations. But it was *how* the white-winged vampires hunted that fascinated me.

While observing my captive colony at Cornell University, I'd seen something remarkable. Crawling across the floor of their feeding enclosure like a pair of spiders, the vampires made what I thought was a bold approach to a rather large hen. The bird cocked her head to one side, eyeing the bats. Her beak could have severely injured or even killed them—and I got ready to intervene. One of the vampires stopped a couple of inches beyond pecking distance but the other crept even closer. Then, amazingly, the bat nuzzled against the hen's feathery breast. Instead of becoming alarmed, the bird seemed to relax a bit. The vampire responded by pushing itself deeper into what I would later learn was a sensitive section of feather-free skin called the brood patch. This was a region densely packed with surface blood vessels, where body heat could be efficiently transferred from the hen to her eggs. Later, the brood patch was where chicks snuggled up to warm themselves. As I watched, the hen reacted to the bat by fluffing her feathers, hunkering down, and finally—closing her eyes.

My God, I thought, *these bats have learned to mimic chicks!*

What was most remarkable to me was that in all likelihood chick mimicry wasn't innate behavior written into the vampire's DNA over millions of years. It must have been learned since the arrival of the Europeans and their domesticated fowl. Were vampire bat mothers teaching this behavior to their young?

So enthralled was I at this wonderfully diabolical maneuver (and its implications) that I never noticed that the second vampire had disappeared under the hoodwinked hen's tail feathers—never noticed until several minutes later when a thin trickle appeared on the floor behind the bird. Through the gloom of the darkened enclosure I could see a small puddle forming and I remember that it glistened like red tinsel.

"We should get these poles up," Jumbo said, nudging me into the present with the business end of a ten-foot stretch of bamboo.

We were setting up shop (thirty-foot lengths of monofilament netting, actually) in one of central Trinidad's least populated regions, Guaico Tamana. Earlier we'd passed through several sleepy towns before Jumbo slammed the jeep into a lower gear and turned off the main road.

Basawan Trace was more of a trail than a road, narrow, twisting, and strewn with potholes. We had bumped along, top down, with Jumbo's soca music cutting through the humid August air. The jeep slowed down only once—to avoid squashing a trio of oilbirds sitting in the road. I'd read that these bizarre creatures employed a form of echolocation to navigate the dark caves where they lived and that the early settlers of Trinidad had named them for their rich reserves of oily fat—which burned quite well in lamps. Now they were mainly a tourist attraction, another checkmark on the Life Lists of the thousands of birders who visited Trinidad each year.

I saw little sign of human habitation in the scrubby forest, but eventually Jumbo pulled up beside a pair of simple clapboard houses. Several garden plots had been carved out of the underbrush, and the yards were strewn with an assortment of old tires, tools, and rusted farm implements. I was soon introduced to the owners, Leno Lara and Mala Boris, as well as their wives, kids, and a friendly assortment of family members totaling about ten people. There was a television playing in the Lara house, but Jumbo informed me later that they had neither running water nor electricity and that the TV was running off a generator.

Everyone seemed to know why we were there and the kids gathered round to watch us set up our poles and mist nets around a fruit-laden grapefruit tree. Jumbo knew from experience that chickens and guinea fowl scrambled up into this particular tree each night, roosting in the branches to escape feral cats and other ground predators. But now the birds were getting progressively weaker with each passing night—bled through the same wounds by the same creatures that had inflicted them—until eventually they dropped from the trees, pale and lifeless. Although vampire bats consume only about half their weight in blood each night (roughly a tablespoon), the anticoagulants in their saliva keep the blood of their prey from clotting, long after the bat has flown off. This charnel house ambience tends to put off most people, especially those unfortunate enough to awaken in a pool of their own blood.

Jumbo and I finished up and were invited back to the Boris residence for some refreshment: warm glasses of the local rum. Twilight is fleeting in the tropics and now, twenty minutes after setting up our mist nets in bright sunshine, it was dark enough that we could no longer see our tree from where we sat under a sheet metal awning.

I asked Mr. Boris if vampires had ever bitten their pigs or their milk cow, but he shook his head. "Just lucky, I guess," he said, and I nodded in agreement.

Unlike chickens, most vampire bat prey does not perish from

blood loss. A half-ton cow can stand to lose a lot of tablespoons of blood before finally tipping over. But an open wound in the tropics is a dinner bell, a beacon on a foggy night. To the hordes of aesthetically challenged flies, beetles, and worms (not to mention a virtual encyclopedia of microscopic organisms), a divot-shaped vampire bite is dining room, bedroom, and toilet—all rolled into one. This generally does not bode well for the animal bearing the wound (or its owner). Infection, disease, and death are the likely outcomes.

Far more serious than disease-promoting wounds, however, is the potential transmission of rabies by infected vampire bats. Rabies is a viral disease that systematically destroys the nervous system of its mammalian victims.* Among the dozens of diseases transmitted by blood feeders like mosquitoes, fleas, ticks, and tsetse flies, rabies, which can only be contracted from another mammal, is perhaps the most feared. It is not the most deadly in terms of numbers of victims, nor is it the most grotesque with respect to outcome, but once the infamous symptoms of rabies appear—hydrophobia, loss of muscle function, and dementia—the disease is nearly 100 percent fatal. Historically, vampire-bat-transmitted rabies had been a terrible problem in Trinidad, killing eighty-nine people and thousands of cattle between 1925 and 1935. In 1934 Trinidad's Medical Department instituted its Anti-Rabies Unit. Part of their job was to respond immediately to any report of vampire bat attacks, and as a result thousands of vampire bats had been netted and destroyed. Others were painted with a poisonous paste that would be groomed off later by roost mates, fueling a chain reaction of death within the colony.

Some of the more conservation-minded workers like Jumbo did their best to calm a frightened public that was already bat

*There is *no* truth to the rumor that bats can carry the rabies virus without becoming sick themselves.

phobic. Local superstition told of the existence of human-sized blood feeders called *soucouyants*. These were supposedly old crones that could shed their skin at night and assume the shape of a fiery ball. To protect oneself from attack, homeowners would sprinkle a bag of rice outside their door. For some reason, the *soucouyant* couldn't enter until she had counted every rice grain.

Rabies control personnel like Jumbo's supervisor, Farouk Muradali, ignored the myths (and I could never envision Jumbo wasting all that rice). Instead, they stressed that only two of the fifty-eight bat species on their island were vampires, and generally speaking, only one of those (the common vampire bat) was a significant rabies threat.

After chatting with the homeowners for about an hour and a half, we checked the mist nets. In one net we had captured a fruit bat *(Carollia)*, and a tiny nectar feeder *(Glossophaga)*. Moving to the second net my flashlight beam illuminated three dark figures. I could see that they were far more muscular than the bats we'd just released and they twisted in the nets, biting and screeching as we approached.

"Diaemus youngi," I exclaimed, donning a pair of thick leather gloves.

"Dey look hungry," Jumbo replied. "And speaking of food . . ."

The vampire bats were carefully extracted from the nets and placed into small cotton bags where they calmed down immediately. A week later they would be among eight specimens of *Diaemus* exported to New Mexico, where they quickly acclimated to the blood of American chickens. The bats' arrival there would spark a minor media frenzy ("Rare Vampires Dodge Death in Desert Town," "Vampire Bats Form Colony in New Mexico") that

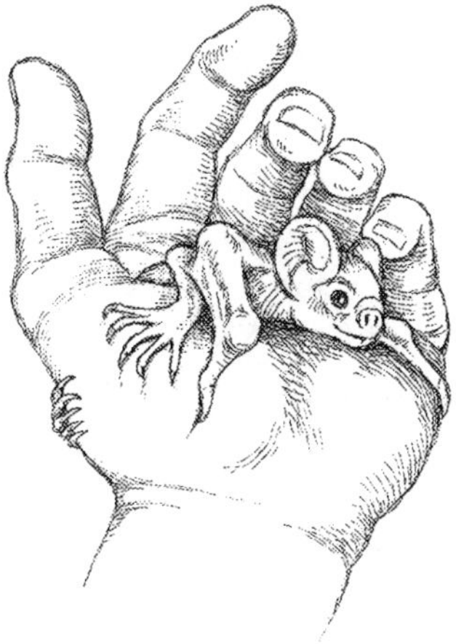

would resurface several months later ("Birth of a Vampire!") when one of the captives delivered a female pup. After a contest publicized by the Long Island paper *Newsday,* the baby vampire bat would be christened Amelia (after another famous female flier).

Jumbo and I stayed out for another hour that humid night in Trinidad, but when the full moon rose we knew there would be no more captures. Vampire bats are notoriously lunar phobic, as are many other bat species.

Two hours later we were eating chicken dinners at an all-night KFC knockoff in downtown Arima.

It seemed like the right thing to do.

As you might have guessed by now, this is a book about blood-feeding creatures and, by association, the substance that they feed upon.* Some of the creatures you'll be reading about, like leeches, bed bugs, and white-winged vampire bats, are mere nuisances. Others—fleas, chiggers, and yes, even the common vampire bat—can be killers. They carry and transmit some of the world's deadliest diseases, including bubonic plague, scrub typhus, and rabies. Still others spread debilitating diseases like Lyme disease and Rocky Mountain spotted fever. And even when they don't transmit disease, fear of these creatures can lead to delusional parasitosis, a condition in which the victim believes that tiny biting or bloodsucking creatures are crawling over his or her body. This is an all-too-common occurrence for those who have experienced a bed bug infestation—or who live in fear of one.

Then there are the truly bizarre sanguivores—blood-feeding finches and vampire moths. And, of course, there's the candiru—a tiny Amazonian catfish whose reported habit of swimming up the human urethra makes it far more feared by locals and tourists alike than its notorious river-mate, the piranha.

Here are the blood feeders—their stories, their strange feeding habits, and the often-devastating effects they can have on the humans they count as food.

This might get a little rough, so grab a glass of red wine and let's get started. . . .

*Vampires are alternatively described as "sanguivorous" or "hematophagous."

part 1

No Country for Old Chickens

We were somewhere around Barstow on the edge of the desert when the drugs began to take hold. I remember saying something like "I feel a bit lightheaded; maybe you should drive . . ." And suddenly there was a terrible roar all around us and the sky was full of what looked liked huge bats, all swooping and screeching and diving around the car, which was going around a hundred miles an hour with the top down to Las Vegas. And a voice was screaming: "Holy Jesus! What are these goddamn animals?"

—Dr. Hunter S. Thompson

1.

WALLERFIELD

(Nine years earlier)

The ceiling tiles in the abandoned icehouse had fallen long ago, transforming the floor of the cavernous building into a debris-strewn obstacle course.

"Hey, it's squishy," I said, stepping gingerly onto a slime-coated chunk just inside the doorway. "Some sort of foam."

"It's probably just asbestos."

My wife, Janet, was a terrific field assistant, but I could tell that this place was already giving her a serious case of the creeps.

"Yes, but with a protective coating of bat shit," I added, trying to cheer her up. "Let's check it out."

Wallerfield, in north-central Trinidad, had been a center for

American military operations in the southern Atlantic during World War II. The land on which it had been built became part of the same Lend-Lease program that had brought Churchill's shell-shocked government fifty outdated American destroyers. Once, it had been the largest and busiest air base in the world, but the English were long gone, as were the Yanks (most of them anyway), and now Wallerfield was an overgrown ruin. Row upon row of prefab buildings had either been carted off in pieces by the locals or reclaimed by the scrubby forests of Trinidad's Central Plain, but because of its cement construction the icehouse was one of the few buildings still standing. Stark white below a mantle of tangled green, the icehouse belonged to the bats—tens of thousands of them.

With help from the Trinidad's Ministry of Agriculture we'd been collecting vampire bats around the island for nearly two weeks—and things had gone incredibly well. So well, in fact, that when our friend Farouk suggested that we visit the cavernous and somewhat notorious ruins of Wallerfield, Janet and I jumped at the chance to accompany him.*

The icehouse wasn't completely dark yet. Daylight streamed through a window frame that in all likelihood hadn't held glass in fifty years. The light fell obliquely onto the floor, illuminating the base of a cement pillar that rose a dozen feet to the ceiling. The only movement was from the dust that swirled into and out of the sunlight. We passed single file through a shaft of motes before continuing on into deepening shadow. The room we were crossing was huge, perhaps two hundred feet long and half as wide, and it took us a good five minutes to pick our way across the slippery rubble.

*Perhaps because of its isolated location, there have been a number of serious crimes associated with this undeniably spooky site (which the American military referred to as Waller Field). In any event, visiting the ruins at night or alone is not recommended.

We stopped at what looked to be a high doorway leading into a smaller room, around fifteen feet square. But instead of entering, our companion put his arm out, stopping us before we could go farther.

"You *don't* want to walk in there, boy." The Indo-Trini accent belonged to Farouk Muradali, head of his government's Anti-Rabies Unit. Farouk would also become my mentor for all things related to Trinidadian bats and a collaborator on a project to study quadrupedal locomotion in vampire bats.

"Why's that, Farouk?" I asked, as Janet and I flicked on our headlamps.

"That is not a room," he said.

As I trained my beam inside the chamber I couldn't help noticing that the floor had a weird shine to it. "What the—?"

"It's an elevator shaft."

"A what?" Janet said, pulling up beside me.

I kicked in a small piece of debris past the threshold and it hit the dark surface with a plop. "Jesus, it's completely filled with water!"

Janet edged closer, the light from her headlamp focused at a point just beyond the doorway. "That is *not* water," she said.

The "floor" of the shaft was a debris-strewn swamp. There was indeed some type of filthy, tar black liquid filling the shaft, but Janet was right—it certainly wasn't water.*

Scattered across the surface of this scuzzy brew were tattered blocks of dark-stained ceiling material as well as unidentifiable rubbish that had been chucked in over the past fifty years. The scariest thing to me was that all of it looked *remarkably* like the rubble-littered cement floor we were currently standing on.

"A group came in here to see the bats some time ago and one of

*I would later learn that the elevator shaft was filled with a combination of bat urine, guano, and rainwater.

them, a woman, turned up missing." Farouk pointed to a spot near where the *real* floor ended. "They found her there, clutching onto the ledge. Only her head and arms were above the surface."

I could see my wife give a shudder and she took several steps back from the edge.

Carefully, I moved a bit closer, kneeling at the entrance of the shaft. It still looked like a solid surface. "Farouk. How deep is this friggin' thing?"

"It goes down several floors," he said, a bit too matter-of-factly. "And off the main shaft—a maze of side tunnels."

As the light from my headlamp moved across the glistening surface, something the size of a football catapulted itself through the beam. My reflexes sent me backward onto my butt as the object landed with a loud splash. Three headlamp beams hit the impact point, but by then whatever it was had disappeared below the ink black sludge.

"What the hell was that?" Janet asked, her voice an alarmed whisper.

"I *think* it was a toad," I responded. "A big mother." And as I turned back to Farouk, he nodded in agreement.

"They feed on the bats that fall in from above," he said. "The babies and the weak ones."

With that, the Trinidadian directed his light upward, until we could just make out the ceiling of the elevator shaft, twenty feet from where we stood.

As I squinted into the darkness, Farouk moved away, motioning us to follow. "You can see the bats much better from upstairs."

Our companion stopped before a narrow stairway leading to the second floor. The railings had either collapsed long ago or been carted off by the locals, leaving only small circular holes in the cement. Three separate beams moved across the steps, each of us searching for any indication that the stairs might not be safe.

I was on the verge of saying something about the strong smell of ammonia when I heard Farouk's voice. His tone had grown more serious. "Janet, maybe you should remain down here."

"Yeah, that's gonna happen," I said with a laugh. My wife had recently spent three hours exploring Caura Cave, the floor of which was slick with guano and crawling with enormous roaches, all without a complaint. Only later did I learn that she had had a migraine the entire time. So it came as no shock when she politely waved off Farouk's chivalrous suggestion and began climbing the darkened stairs.

One year earlier, at a symposium on bat research, I had gotten up the courage to approach Arthur M. Greenhall, one of the world's leading authorities on vampire bats. I was in the second year of a Ph.D. program at Cornell and like many grad students I was sniffing around for a dissertation project. (Luckily, the head of my graduate committee, John Hermanson, wasn't one of those guys who handed you a ready-made project, although I had to admit there were some days when I wished he had.) By this time, Greenhall was in his midseventies but he was still vibrant and inquisitive—as excited about science as anyone I had ever met.

Born and raised in New York City, he'd had a storied career. In 1933 Greenhall and Raymond Ditmars, his mentor at the New York Zoological Park, had collected the first vampire bat ever to be exhibited alive in the United States. It was a female that turned out to be pregnant, delivering a vampire bat pup several months later. The following year, the young scientist arrived in Trinidad during the height of a major rabies outbreak. He studied the deadly virus and its blood-feeding vector with local scientists and collected additional vampire bats. On his return to the United

States, he found he had more specimens than his zoo could display or handle. Greenhall solved the problem by keeping twenty of the creatures in his New York City apartment for two years.

During a break between research presentations that day, I had spoken to several noted bat biologists about possible differences in behavior or anatomy between the three vampire bat genera, *Desmodus, Diaemus,* and *Diphylla.* From previous studies I had learned that *Desmodus,* the common vampire bat, exhibited an incredible array of unbatlike behaviors, including a spiderlike agility on the ground. Just as interesting to me was the way *Desmodus* initiated flight. In virtually all nonvampire bats, takeoff began with a wing beat that accelerated the animal away from the wall, ceiling, or branch from which it hung. Heavily loaded down after a blood meal, *Desmodus* was renowned for its ability to catapult itself into flight from the ground by doing a sort of super push-up.

"Maybe," I proposed, "the other vampire bats, *Diaemus* or *Diphylla,* did things a little differently."

"Not likely," I was told more than once.

"A vampire bat is a vampire bat is a vampire bat," chanted several bat scientists. I wondered if there might be a secret handshake that went along with this information, one that I had yet to learn.

After introducing myself to Greenhall, I told him what the bat researchers had said, adding that I found their responses puzzling.

"Why's that?" the vampire maven responded.

"Well, because the rule of competitive exclusion says that if similar animals are competing for the same resource, in this instance blood, then one of three things will happen. One of the animals will relocate. One of them will go extinct. Or one of them will evolve changes, reducing the competition for that resource."

"And since vampire bat genera have overlapping ranges . . . ?" Greenhall interjected, setting me up beautifully for the punch line.

"They've *got* to be different."

The old scientist gave me a sly smile. "You're on to something, kid," he said. Then he lowered his voice. "Now get on the stick before someone else gets to it first."

It had taken me six months to "get on the stick," but by then my fellow Cornell grad students, Young-Hui Chang and Dennis Cullinane, and I had followed our mentor John Bertram's lead and built a miniature version of a force platform, a device that could measure the forces applied to a flat metal plate as a creature (in this case, a vampire bat) moved across it. By synchronizing the force platform signals with high-speed cinematography, we planned to see if there would be measurable differences in the flight-initiating jumps of *Desmodus rotundus* and *Diaemus youngi*, the two vampires I would collect and bring back from Trinidad.

Not long after arriving in Trinidad and Tobago's capital, Port of Spain, I told Farouk what a pain it had been for us to machine the metal components of our force platform, get the electronics working just right, and then write the data-acquisition software. He stood by patiently as I tooted my own horn, polished it a bit, then tooted some more. Finally, I ran out of intricate gear to describe (or it might have been air).

"It won't work," Farouk said, matter-of-factly.

"Excuse me?" I replied, my voice cracking like a twelve-year-old boy's.

"Your experiment won't work."

Now I was getting visibly annoyed. Hadn't I just told him how much time, effort, and brainpower had gone into this project?

"Of course it'll work." I was getting frantic now.

The Trinidadian said nothing.

"Why won't it work?"

Muradali put his hand on my shoulder and smiled. "Because *Diaemus youngi* doesn't jump."

"Oh," I replied, sheepishly. "Right."

The light from Janet's headlamp swept upward from the bottom of the empty elevator shaft (now below us) to the ceiling. "So where are all the—" Her beam had stopped tracking abruptly.

Illuminated at the top of the chamber were three circular clusters, each composed of a dozen or so black silhouettes, arranged concentrically. They hung silently, reminding me of giant Christmas tree ornaments. Suddenly, one of the fusiform shapes unfurled, revealing wings nearly two feet across.

"Phyllostomus hastatus," Farouk whispered. "The second-largest bat in Trinidad."

"Crawling mother of Waldo," I muttered, and Muradali threw me a confused look.

"Don't mind him," Janet explained, keeping her light trained on the bats. "He gets all scientific when he's excited."

Muradali nodded politely, then began assembling an object that looked suspiciously like a drawstring-equipped butterfly net at the end of a four-foot pole.

I shot him a quizzical look. "A butterfly net?"

"Swoop net," Muradali corrected, handing it to Janet.

Farouk nodded toward the net, then shined his light up at a cluster of bats. "To catch the ones closest to the elevator door, you lean out over the edge while someone holds your belt or backpack."

Janet glanced up at the bats, then quickly shoved the net into my hands. Possibly she'd had the same vision that I'd just had, of tumbling down a concrete-lined abyss with nothing except years of rainwater, bat guano, and asbestos to soften the fall.

As I moved into the doorway, it was impossible to chase away the image of that poor woman, stepping off the solid concrete floor and into a bottomless pit of bat-shit soup. "Thanks, hon," I said.

Janet only smiled.

"We'll leave these bats alone," Muradali said, moving away from the shaft.

As we quickly followed him, I let out a breath I hadn't realized I'd been holding. "Can we catch vampires like this?" I asked, suddenly feeling a bit braver and taking a few swings at some phantom air bats.

"No," he replied, picking his way through the debris. "Too smart."

Later, the scientist explained that early efforts to eradicate vampire bats had resulted in the deaths of thousands of non-blood-feeding species. In 1941, Captain Lloyd Gates was placed in charge of protecting the American forces stationed at Wallerfield from the twin threat of mosquitoes and vampire bats. Gates's less-than-subtle response to the bat problem was to have his men use dynamite and poison gas in caves known to contain bat roosts. Flamethrowers became a popular alternative, but still the vampires persisted, as did their attacks upon the encroaching military men. Also hard hit was the increasing population of locals who had been drawn to the region for the income the base provided. As a result, thousands upon thousands of non-blood-feeding bats were blown up, poisoned, or incinerated. Even worse, these bat eradication techniques were apparently so appealing that over eight thousand caves in post–World War II Brazil were similarly destroyed.*

Farouk recounted how he and vampire bat expert Rexford Lord had been sent to Brazil to pick up some tips on eradicating *Desmodus* from the antirabies groups working there.

*Art Greenhall told me that the same grim methods had been employed in Venezuela, where nearly a million bats were killed annually from 1964 through 1966.

"These guys took us to a cave. Then they rolled out a big tank of propane and wired it up with an old-fashioned camera flash, running the wires out the cave entrance."

He described how everyone waited outside the cave entrance while one of the Brazilians opened up the gas-tank valve.

"Must have been the new guy," I added.

"They used a triggering box to set off the flashbulb and the explosion ripped through the cave like a bomb," Farouk said. Then he shook his head and continued. "After the smoke cleared, they asked us to go in and identify the dead bats that we found. And there were *thousands*. All sorts of species—but not one vampire."

Farouk said that later on the men ventured deeper into the cave and there, lined up above a ledge, was a row of dark shapes.

"They were vampire bats. All of them were looking quite fit and not at all disturbed by the explosion. The bats that died in there were a lot more delicate."

The Brazilian cave fiasco hadn't solved the vampire bat problem, but it did serve to illustrate how *Desmodus* had evolved to become extremely opportunistic, extremely intelligent, and extremely difficult to eliminate.

At this point Farouk got to the heart of the matter. "Feeding on blood is a tough way to make a living."

Back at Wallerfield, we moved deeper into the building, using our headlamps to avoid tripping over the ceiling, a concept I was just beginning to wrap my mind around. The acrid ammonia smell was getting even stronger and suddenly we were in Bat Central.

The lights and our movements had finally aroused the aerial residents of the icehouse and now there were hundreds of furry bodies flashing past, their barely discernible high-frequency calls set against the parchment flutter of wings.

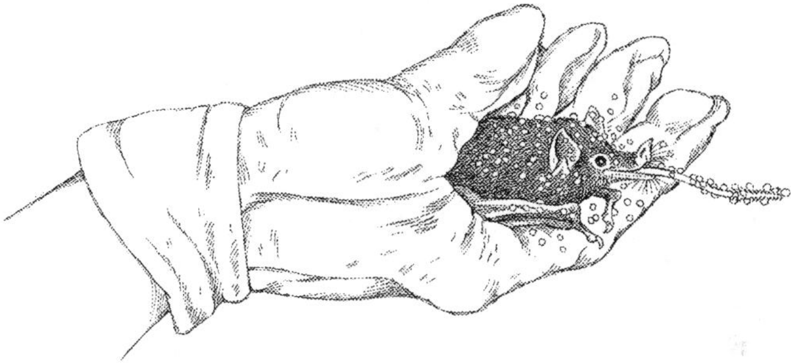

I turned off my headlamp and took a couple of swings with the swoop net. Almost immediately I felt a slight difference in the weight of the net and tugged the drawstring tight.

I flicked my light back on. Reaching in a gloved hand, I plucked out a tiny struggling form, manipulating it gently so that the wings were folded and pinned against the body. A struggling animal, no matter how large or small, was far more apt to hurt itself, and the person handling it, if it wasn't fully and comfortably restrained.

Janet and Farouk pulled in close, focusing their headlamp beams on my delicate captive. The bat had an extended snout and a long, protractible tongue that seemed to be equipped with a brushlike tip. Its teeth were tiny and weak and the creature soon gave up trying to bite through my leather batting gloves.*

*Lightweight and perfect for handling small flying mammals or moving through a thorn-laden forest, I'm still amazed that some people cling to the belief that these gloves were named for America's national pastime.

"Glossophaga soricina," Farouk said. "A nectar feeder."

The bat looked as if it had been assaulted by a powder puff. The "powder" was actually pollen that the creature had inadvertently picked up while feeding. Like hummingbirds, *Glossophaga* and their relatives were vital components of their ecosystems, in fact, over five hundred species of tropical plants were at least partially dependent on bats to pollinate them.

The nectar-feeding lifestyle was also a great example of convergent evolution, in which organisms (in this case several dozen bat species and over three hundred species of hummingbirds) evolved to resemble one another (anatomically and behaviorally), not because they were closely related but because they existed in similar environments or exploited a similar resource. In this instance, the resource was nectar, the sugar-jacked liquid produced by many plants with an evolutionary ulterior motive. While obtaining its meal, this bat (like hummingbirds or insects like bees and butterflies) had been dusted with pollen, pollen that would now be delivered via airmail to some fertile and, quite possibly, distant flower. It was a coevolutionary relationship that had been going on since the flowering plants first evolved during the reign of the dinosaurs.*

Additionally, just as in other examples of evolutionary convergence, there *were* major differences between bat and bird pollinators, and some of these (beyond the obvious daytime- vs. nighttime-feeding habits) were quite significant. For example, hummingbirds, which number around 340 species, are renowned for their ability to hover for extended periods as they feed. Re-

*Fossil evidence indicates that insects may have forged a relationship with flowering plants (angiosperms) soon after the latter appeared, some 120–130 million years ago. The first bats (which were insect eaters), as well as the ancestors of modern hummingbirds, appear to have evolved around the time that the nonavian dinosaurs (and significantly, their flying cousins, the pterosaurs) went extinct, around 65 million years ago. With pterosaurs no longer filling the aerial vertebrate niches, birds and bats underwent a rapid diversification.

markably, they accomplish this maneuver with wing-beat frequencies that can approach ninety beats per second. On the other hand, those relatively few bat species that can hover (certainly fewer than twenty), generally do so for less than a second with wings that max out at around twenty beats per second.

Another difference between bat and bird pollinators concerns the upstroke portion of the wing beat. All bats use the same muscles to raise their wings that humans use to extend their arms out to the side. In both bats and humans, these muscles (i.e., the deltoid and supraspinatus) extend from the back of the shoulder (the scapula) and attach to the upper arm bone (the humerus). When these muscles contract, it's like pulling the strings on a marionette's arms—but with the power to lift the wings coming from muscle contraction rather than a puppeteer.

In terms of flight efficiency, though, the important factor is that in bats the upstroke muscles are located *above* the wing. Since it is

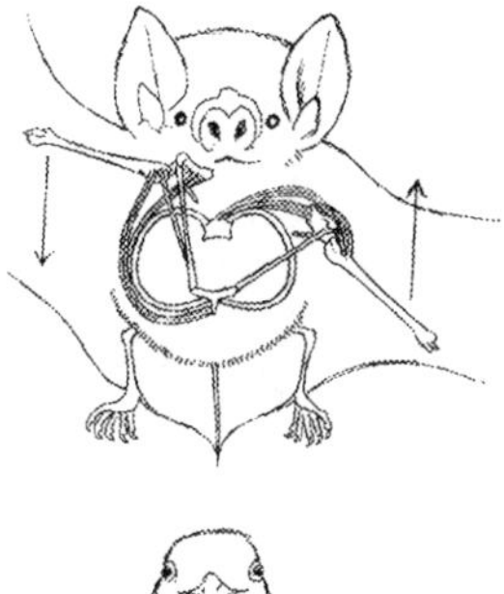

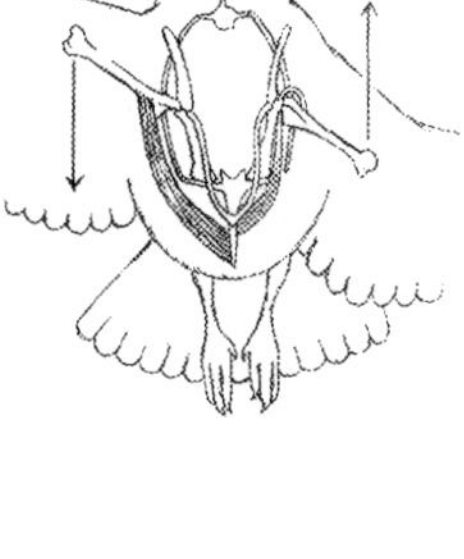

more aerodynamically efficient to have as much weight as possible *below* the wing, this extra weight reduces flight efficiency, giving bats their characteristic flittery flight.*

Birds have evolved a solution to this problem since both their downstroke and upstroke muscles are located *below* the wing. Situated on the sternum (deep to the bird's downstroke-driving pectoral muscles), the supracoracoideus muscle sends its long tendon snaking through a hole in the shoulder joint to an attachment site on the humerus. When the supracoracoideus muscle contracts, its tendon acts like a pulley to raise the wing. The end result is a smoother (less jerky) flight in birds compared to bats.

These performance differences follow a general trend in most flight characteristics in which birds are more aerodynamically efficient than bats. This is almost certainly because birds have been flying (and, in the case of hummingbirds, hovering and feeding on nectar) far longer than their mammalian counterparts.

Back at Wallerfield, Farouk nodded at my tiny captive. "You should release that *Glossophaga* before we leave," he said. "If you want it to live."

"Why's that?" Janet asked. We'd been bagging bats in Trinidad for several weeks, then taking them back to the PAX Guest House where we were staying in Tunapuna.†

*Think of where the baggage and cargo are stored in an airplane, or alternately, how no one fights to get the turkey's back at Thanksgiving dinner.

†PAX sits perched atop a hill overlooking the Caroni Plain, and is located on the grounds of a Benedictine monastery. Our friends there, Gerard Ramsawak and his lovely wife, Oda, had set up a wonderfully serviceable lab for us in what doubled for a garage. After recording a series of measurements and tracing wing shapes, Janet and I would wait until dark before releasing the bats into the night.

"Glossophaga has a very high metabolic rate," Farouk replied. "If that one doesn't get nectar tonight, it will starve to death."

"Yikes," I said, glancing down at the bat with renewed interest.

Janet nudged my arm. "Sounds like those shrews we caught with Deedra and Darrin last year at Arnot Forest."

Janet had nailed it. Shrews are tiny, insectivorous bundles of energy. Superficially, they resemble rodents (another example of convergent evolution), but they have amped-up, nutrient-burning bodies, that, like the nectar-feeding bats, require a constant and relatively immense intake of energy. The shrews we'd taken during a mammal survey in a forest near Cornell had a resting heart rate of approximately eight hundred beats per minute, and when pressed, they could reach fifteen hundred beats per minute—the highest ever recorded for a mammal. As a consequence, shrews have to eat almost constantly—worms and insects, mostly—but sometimes even other shrews. Their aggressive demeanor and toxic bites also enable them to tackle animals much larger than themselves. During one of our long nights in the field, I'd brought up the topic of a creature feature I recalled seeing as a kid. It was the unintentionally funny, 1959 horror flick, *The Killer Shrews,* in which dogs outfitted with goofy shrew wigs, terrorized a handful of cocktail-guzzling scientists, a well-endowed young woman, and a testosterone-squirting hero in a captain's cap. Besides a last line that rivaled Clark Gable's in *Gone with the Wind,* what I found most memorable about this mostly forgotten cinema "classic" was the fact that the filmmakers had gotten at least one thing right (two, actually, if you count the alcohol intake by the scientists). If indeed shrews *had* evolved or, in this case, mutated, to be the size of dogs (even small dogs)—humans would have had a *serious* and unbelievably vicious predator to contend with. Luckily for those of us collecting real shrews, there was no danger—only the discomfort of late nights during which we had to check over a hundred "live traps" every two hours—to prevent our hyperactive captives from starving to death.

In the icehouse at Wallerfield, Janet and I took a last look at the amazing little pollinator.

"See ya," I said, gently flipping the bat upward.

The tiny creature disappeared in a whisper of parchment.

I looked over at Farouk, who nodded and motioned toward the stairwell. "We'd better get going, Bill. We don't want to be out here after dark."

"Second that," Janet said.

I turned to say something to my wife, but she was already moving toward the exit.

"Right," I said, following the beam from Janet's headlamp as she sought the comfort of sunlight.

Like nectarivory, blood feeding in bats is another highly specialized lifestyle, but there is little or no convergence between birds and bats, in all likelihood because there's no competition between the two groups. While there are birds that regularly feed on blood (e.g., vampire finches and, indirectly, those that pick ectoparasites like ticks off of large mammals), none of these birds is an obligate blood feeder like the vampire bats. In other words, no bird species will starve to death in two or three days if it doesn't secure a blood meal. This means that as far as *vertebrate* sanguivores are concerned, bats hold the exclusive rights to their aerial and terrestrial niches.*

So what did the early naturalists have to say about vampire bats, and how did these creatures become forever tied to the

*There are, however, literally thousands of invertebrates that have evolved to feed solely on blood.

growing vampire hysteria that was simultaneously taking place in Europe? How did blood feeding evolve in bats, and why has it never appeared in birds—an older and more diverse group? Oh, and finally, why is just about everything people *think* they know about vampire bats completely wrong?

It might be best to start with this last question.

Can you tell me why in the Pampas, ay and elsewhere, there are bats that come out at night and open veins of cattle and horses and suck dry their veins; how in some islands of the Western seas there are bats which hang on the trees all day, that those who have seen describe as giant nuts or pods, and that when the sailors sleep on the deck, because that it is hot, flit down on them—and then in the morning are found dead men, white as even Miss Lucy was?

—Bram Stoker

2.

CHILDREN OF THE NIGHT

When the explorers of the New World returned home to Europe in the fifteenth and sixteenth centuries, they were far more concerned with gold, God, and geography than they were with accurate zoological accounts. Amid fanciful tales of sea serpents, giants, and mermaids, there were also reports of bats that fed at night upon the blood of unfortunate humans and their livestock. Although these creatures were generally described as being hideous, with wingspans of up to five feet, nobody actually took the time to figure out which bats were vampires and which weren't. The rule of thumb seemed to be that the largest and ugliest bats were vampires—and, on both accounts, the explorers were dead wrong.

Early taxonomists contributed significantly to the confusion. Carl von Linné (who actually Latinized his own name) and the morphologist Étienne Geoffroy Saint-Hilaire were responsible for initiating a misunderstanding regarding bats and blood feeding that still exists today. With little knowledge of the bat's biology and no regard for their actual diet, they assigned scientific names like *Vampyrum spectrum* (which happens to be a *really* large bat), *Vespertilio vampyrus, Vampyressa,* and *Haematonycteris* to bats that had never so much as snuck a sip of blood.*

Even card-carrying tropical zoologists got things horribly wrong. Johann Baptiste von Spix, curator of the zoology collection at the Bavarian Academy of Sciences, had spent nearly three years on a collecting trip to Brazil starting in 1817. He returned with thousands of specimens, many never before seen in Europe. One of these was *Glossophaga soricina* (the pollen-dusted bat I had "swoop-netted" at Wallerfield). Spix described *Glossophaga* as "a very cruel blood-sucker" *(sanguisuga crudelissima),* hypothesizing that the creature we now know to be a delicate hummingbird mimic actually used its brushlike tongue tip to reopen the wounds it had somehow inflicted with its tiny teeth.

The chiropteran disinformation campaign continued well into the nineteenth century. By this time collectors were swarming all over the Neotropics in an effort to supply the burgeoning museums and private collections of Europe. Even though naturalists like Charles-Marie de La Condamine and Alfred R. Wallace had begun writing more factual accounts of vampire bat attacks, these creatures were still considered to be mythical by many in the European scientific community. The problem was that while the slaughterhouse results of a nighttime vampire bat attack were

*Like other carnivores, *Vampyrum* ingests blood, but not as a sole source of nutrition.

easy enough to record, identifying the actual bat that left the mess was more of a poser. And, as it turned out, even when the culprit was correctly identified, prejudice got in the way.

In 1801, in Paraguay, the Spanish cartographer and naturalist Felix D'Azara collected the creature that would eventually become known as the common vampire bat. But even though D'Azara asserted that this was the bat responsible for attacks on humans and livestock, British and French taxonomists thumbed their noses at his claim. In 1810 the same bat was named and described by Geoffroy. *Desmodus* (literally, "fused tooth") was named for its unique incisors: a chisel-shaped set of uppers and a uniquely bi-lobed pair of lowers. Unfortunately, there was absolutely no mention of blood feeding in Geoffroy's description of *Desmodus*. Similarly, in 1823 Spix named and described a bat that had been collected in Brazil, but it would be years before *Diphylla ecaudata* would be recognized as a second vampire bat species.*

It wasn't until 1832, when Charles Darwin and his servant observed *Desmodus rotundus* feeding on a horse, that the English-speaking world had a name to associate with the blood-feeding deed.

> The vampire bat is often the cause of much trouble by biting the horses on their whithers. The injury is generally not much owing to the loss of blood as to the inflammation which the pressure of the saddle afterwards produces. The whole circumstance has lately been doubted in England; I was therefore fortunate in being present when one was actually caught on a horse's back. We were bivouacking late one evening near Coquimbo, in Chile, when my

*In 1893 the last piece of vampire bat puzzle was completed when the third vampire, *Diaemus youngi*, was identified. By this time scientists had finally figured out that the bat they were describing actually fed on blood.

> servant, noticing that one of the horses was very restive, went to see what was the matter, and, fancying he could detect something, suddenly put his hands on the beast's whithers, and secured the vampire.*
>
> (Charles R. Darwin)

Because of similarities in appearance, behavior, and range (parts of Mexico, the warmer regions of South and Middle Amer-

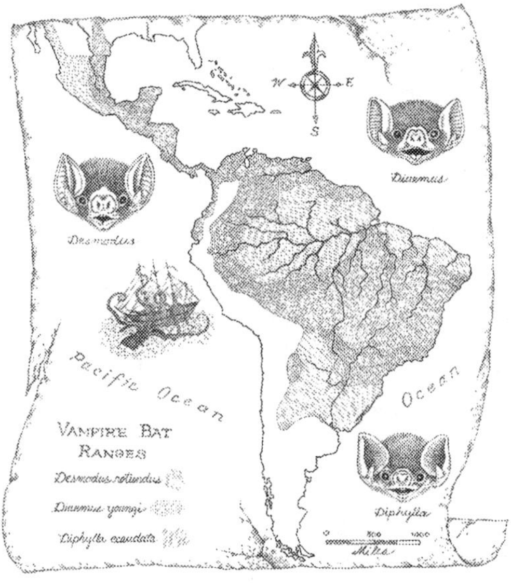

*Admittedly, this is rather bizarre behavior on the part of the common vampire bat *and* its collector. As anyone who has observed these creatures in the field knows, vampire bats are unbelievably secretive—especially, it seems, around humans. Why then, did this particular bat allow itself to be approached by two men, only to be plucked off its host by Darwin's servant? Even had this bat stuck around (and that is doubtful), no one who handles vampire bats would have tried to capture one by hand without first donning a pair of thick leather gloves (which Darwin makes no mention of). That's because *Desmodus rotundus,* the bat described by Darwin, bites ferociously when handled. In the end, I can think of three explanations for this very un-vampire-bat-like behavior: Darwin forgot to mention the gloves, the bat was sick, or Darwin embellished his description of the encounter.

ica, plus the islands of Trinidad and Margarita), *Desmodus, Diaemus,* and *Diphylla* were initially placed into their own family, the Desmodontidae. More recently, researchers have reduced them to a subfamily within the large, primarily Neotropical family Phyllostomidae. There are around one hundred and fifty phyllostomids (i.e., members of the family Phyllostomidae) and they're sometimes referred to as New World leaf-nosed bats. This is because they live in the Americas and most of them have a vertically projecting, spear-shaped nasal structure. Although nose leaves may look menacing, they are actually soft and pliable.

Early naturalists claimed that nose leaves were used by vampire bats as deadly flesh stilettos, to gouge victims before a blood meal. Many years later, scientists studying the strange ultrasonic capabilities of bats uncovered an interesting, though decidedly less gory function for the nasal protuberances. Just as a megaphone can be used to direct the human voice, the nose leaf is

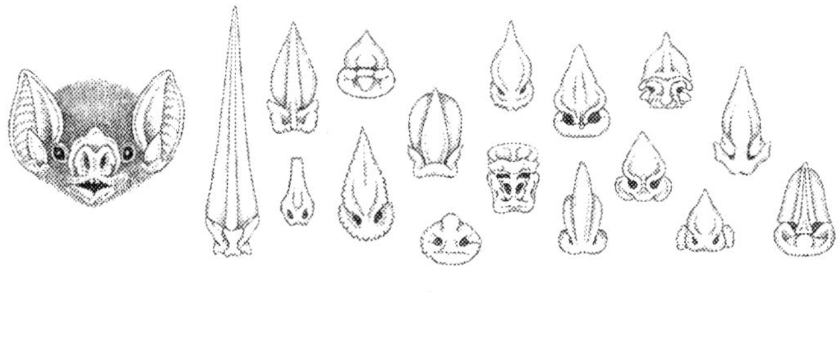

actually involved in directing the echolocation calls emitted by the bat. Ironically, nose leaves are greatly reduced in size in vampire bats (like *Desmodus*) where they function primarily in thermoperception—the ability to sense differences in temperature. This is an adaptation that comes in handy as vampire bats approach their warm-blooded prey in complete darkness. Once the bat gets within around fifteen centimeters of its target, thermoreceptors in the low, ridgelike nose leaf can detect the slight temperature differences that exist in areas of the skin where blood vessels lie just below the surface. The bat uses this information to help determine where a bite will be made.*

In hindsight, the function of the bat nose leaf was one more bit of misinterpreted information for early naturalists, who used the presence of this structure to mistakenly categorize over a hundred species of non-blood-feeding bats (e.g., *Glossophaga*) as vampires. Along these lines, it should also be noted that nose leaves occur in two additional (and only distantly related) families of Old World bats, the Rhinolophidae and the Megadermatidae (the latter is now commonly known as false vampire bats). This is yet another example of convergent evolution, and although neither of these groups have any blood-feeding members, the presence of a nose leaf probably contributed to claims of vampire bats inhabiting Europe, Africa, Southeast Asia, and the Indo-Pacific.†

*A recent study suggests that *Desmodus* uses passive hearing to identify the breathing patterns of animals that have been fed upon previously. This may help explain how vampires can return to the same animal over consecutive nights.

†In 1967 a man fired a shotgun into a cluster of bats in an abandoned railway tunnel west of Comstock, Texas (and about five miles from the Mexican border). One of the dead bats was determined to be *Diphylla ecaudata*. Over forty years later, this remains the only modern record of a noncaptive vampire bat in the United States.

Even though the identity of the three vampire bats was not fully known until the 1890s, bloodletting bats have been referred to as vampires since the mid-1700s, and although vampire folklore did not begin with the discovery of vampire bats, it was clearly strengthened by it.

According to folklorist Stu Burns, the word *vampire* has its somewhat hazy roots in the Slavic proper name *Upir*, first recorded in an eleventh-century Russian manuscript. *Vampire* (or *vampyre*—used hereafter to denote the mythical bloodsucker) is a westernization of *Upir* (or *Upyr*) and the word appears to have been coined in English in a pair of 1732 publications. *Vampyre* refers to a corpse that has returned from the dead to drink the blood of the living. Similar creatures were said to haunt the rural villages of nearly every Slavic nation. Not surprisingly, each culture gave their monster its own name (e.g., *vukodlak* in Serbia, *strigoii* in Romania, *eretika* in Russia, *insurance salesman* in . . . well, never mind).

It should also be noted that stories of vampyrelike creatures have a worldwide distribution. Bloodsuckers inhabit the folklore and literature of ancient China, Babylonia, and Greece, as well as the pre-Columbian cultures of Mesoamerica (most notably as the Mayan bat god Zotz or Camazotz).

Vampyre hysteria ebbed and flowed throughout Europe in the fifteenth and sixteenth centuries, reaching its peak in the 1730s. At this time it became quite popular to dig up dead bodies, accuse them of crimes, and then smash a stake through their decaying hearts. According to legend, those corpses hoping to avoid skewering often chose to transform themselves into something not quite so corpselike. Although Slavic vampyres never actually took the form of bats, popular transformation destinations included animals or inanimate objects such as fire and smoke. Fear was an important component of most vampyre legends, but some of these creatures would have had a hard time striking terror into your average toddler. For example, Muslim gypsies in the Balkans won't keep pumpkins or watermelons for more than ten days (or after Christmas) for fear that they'll transform into vampyres. Thankfully, these vampyre veggies have no teeth—so they're reduced to pestering people by rolling around the ground, growling, and dripping blood.

Descriptions of how vampyres attacked their prey are almost completely absent from the early folklore, but there is some general agreement that previously healthy victims began wasting away before ultimately succumbing to the vampyre's supernatural powers.

Some scholars have attempted to explain the multicultural obsession with vampyrism from a criminal standpoint—as gruesome acts committed by individuals exhibiting actual medical conditions ranging from schizophrenia to rabies. On rare but memorable occasions, criminals turned up who were actually obsessed with blood. These "vampyrists" were psychotic rather than supernatural, obtaining gratification by consuming or otherwise coming into physical contact with the blood of others. The most infamous vampyrist may have been the Hungarian countess Elizabeth Báthory. Apparently the countess was quite fond of brutalizing her servants, and after slapping one young woman in the face, she found herself splattered with the girl's blood. Soon after, Báthory became convinced that the liquid had cosmetic and restorative powers. Ultimately, she may have participated in the torture and murder of over six hundred maidens—all of this mayhem so that she might drink or bathe in their blood.* After her trial in 1611, the countess was walled up inside a small chamber within her own castle where she lived out the last three years of life in darkness and solitude. In what might have been an early attempt at a plea bargain, several of Báthory's assistants avoided similar confinement by having their extremities hacked off and then getting burned at the stake.

Some researchers seeking to explain our fascination with the vampyre phenomenon looked to the deaths themselves rather than the crimes surrounding them. They related fatalities that resulted from supposed vampyre attacks to diseases like anemia, tuberculosis, or the various plagues (such as the Black Death) that spread in wavelike fashion across Europe and much of the globe.† Additionally, given the general population's ignorance about medical conditions like comas, it's no shock that there were numerous reports of what may have been premature burials and encounters

*One wonders just how she got around the problem of blood coagulation.

†In Victorian England, people consumed blood at slaughterhouses, convinced that by doing so they could prevent tuberculosis, a deadly bacterial infection, once referred to as "consumption" (not because of blood-drinking but because it seemed to consume people from within).

with "dead" people who had suddenly and inexplicably come back to life.*

Clearly, though, once word of the existence of real vampire bats began to circulate, a new supernatural emphasis on these mysterious (and as yet unidentified) creatures began to take shape. Bats living in Europe, where blood-feeding species had *never* existed, were gradually implicated as being vampyres. Hysteria and storytelling outpaced reason and science (although to be frank, science had done a lousy job of getting its vampire bat stories straight). Gradually, the folklore of vampyrism began to incorporate the bat and batlike characteristics into its lexicon. Unlike the birds, bats were mysterious and barely glimpsed creatures of the night; they resembled rodents (at least superficially) and flew on leathery wings. Bats were prime candidates for superstition and unwarranted fear, and they would become forever linked to vampyrism in 1897 with the publication of Bram Stoker's novel, *Dracula.*

Inspired perhaps by similar stories about how Mary Shelley and Robert Louis Stevenson had come up with their ideas, Stoker (an Irish theater manager and critic) joked that the literary inspiration for his most famous work came from a nightmarish dream that followed a late evening meal of dressed crabs.

Stoker derived the title of his novel from a real-life, fifteenth-century Romanian *voivode* (warlord or prince). Vlad III, from the principality of Wallachia, became infamous for the means by which he slaughtered his primarily Muslim enemies. Although he

*A boy by the name of Ernest Wicks was found, apparently dead, in 1895, and after being laid out at a Regent's Park mortuary, he suddenly returned to life. An investigation later revealed that the child had "died" four times and that "the mother has obtained no less than three medical certificates of death, any one of which would have been sufficient for the subject to have been buried." As we'll see later, George Washington's final request was that he not be entombed for three days after being pronounced dead—presumably for fear of being interred alive.

utilized a wide variety of tortures ("He blinded, strangled, hanged, burned, boiled, skinned, roasted, hacked, nailed, buried alive, and ... stabbed"), Vlad's favorite torture method was to have victims impaled through the heart, chest, or navel on sharpened wooden stakes. Mothers were stabbed through their breasts before having their babies thrust onto the jagged shafts. In other instances, victims were pierced from the buttocks, upward, by a stake that had been rounded off and lubricated to prevent the impaled from dying too soon.

Slaughtering on a massive scale, the prince reportedly covered the landscape with thousands of staked bodies in various stages of decay. These "forests of the impaled" instilled fear in Vlad's enemies and eventually earned him the moniker Vlad Tepes (Vlad the Impaler).

How did a murderous Romanian prince lead Bram Stoker to his famous title? It's quite simple. Vlad's father (Vlad II), who was also a prince, had been indoctrinated into the Order of the

Dragon* around 1431 and was thereafter known as Vlad Dracul. Those who knew Vlad the younger could avoid the embarrassing "Impaler" title by instead referring to their prince as Dracula—literally, "son of the dragon." It should also be noted that since Dracul has a dual meaning in the Romanian language—"dragon" and "devil"—some people have interpreted the name Dracula in a more sinister light.

Even after establishing a link between vampires and vampyres, questions remain about the real-life creatures—questions that have puzzled and intrigued those of us who study them: How did blood feeding evolve in vampire bats? And why (among twenty thousand species of terrestrial vertebrates) is obligate vampirism confined to only three, closely related New World bats?

First of all, as far as the origin of vampire bats is concerned, the fossil record (so important in detailing the life histories of many prehistoric creatures) is no help here. Although there are several species of fossil vampire bats (including a supersized version, the wonderfully named *Desmodus draculae*), these bats are clearly vampires, not transitional forms that might shed light on their previous feeding habits. Paleontologists get all tingly at the very mention of transitional forms. But to better understand them, let's leave vampire bats for a minute to examine what is arguably the most famous of these transitions—one that beautifully illustrates the evolutionary changes that led to the modern horse.

Using the combined results of both classical and modern studies, vertebrate paleontologists have been able to correlate gradual changes in the skull, teeth, and limbs of horse ancestors with en-

*This group of selected nobles (friends and political allies of the Holy Roman Emperor) had been commissioned in 1408.

vironmental changes that took place on the North American continent starting some fifty million years ago (during the early Eocene epoch). One of the groups that evolved to fill the niches left open by the dinosaurs was a rather diverse assemblage of mammals called the Perissodactyla (odd-toed ungulates).* Within this group, which also included the ancestors of rhinos and tapirs, was *Hyracotherium*, a fox-sized creature that inhabited the extensive forests that covered much of the region. With short legs and eyes set in back of a short snout, *Hyracotherium* was well adapted for a life spent hiding in the underbrush and browsing on soft, leafy plants and fruit.

Starting around twenty-five million years ago (as indicated by clues such as changes in fossil plant species and their seeds), the climate in North America gradually became drier. Forests dwindled and grasslands spread. Some of the small browsing mammals went extinct (as did many other forest types), but others survived, mainly because they evolved adaptations for coping with their new environments. For example, higher crowned (i.e., longer) teeth enabled these mammals to deal with the constant wear and tear of eating the tough, silica-laden† grass that had replaced the soft leaves and shoots popular with forest diners.

With less plant cover in which to hide, longer limbs became important for moving quickly over open ground. Basically there are only two ways to augment running speed: by increasing stride frequency and by increasing stride length. Longer limbs contributed to the latter since each time the limb moves forward during a stride, more ground is covered. As the limbs lengthened, toes that were once on the ground either disappeared or remained as

*Ungulates are hoofed mammals. Artiodactyls are ungulates (like cows, camels, giraffes, and pigs), so-named because they have an even number of toes (i.e., two or four).

†Silica is also a primary component of glass (which is also rather indigestible).

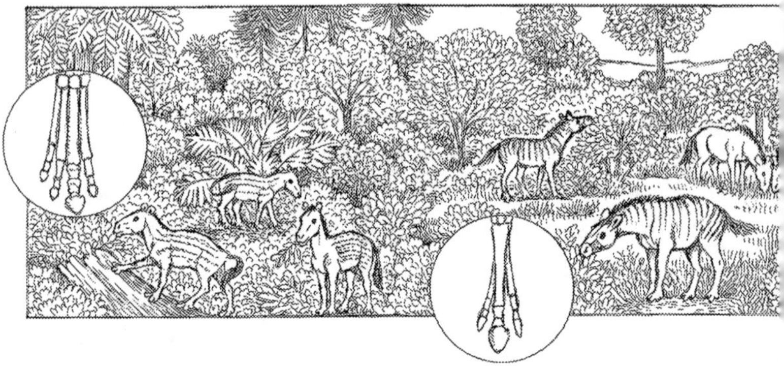

vestiges, like the splint bones found in the front legs of the modern horse, *Equus cabalis*.*

Protohorse skulls became longer as well, with the eyes set farther back from the mouth. Longer snouts (rostrums) allowed these creatures to graze while simultaneously watching for predators.

In addition to looking more and more like modern horses, these ungulates became extremely diverse—with up to fifteen North American species living at the same time (around ten million years ago). For whatever reasons, though, by roughly five million years ago, only the modern horse remained, spreading into Asia and Europe across a land bridge that spanned what is now the Bering Strait (separating Russia from Alaska). By about thirteen thousand years ago, climate changes, humans, or perhaps, as hypothesized by American Museum of Natural History Curator of Mammalogy Ross MacPhee, a rabieslike hyperdisease drove many large North

*The modern horse runs on the tips of its third (or middle) digit.

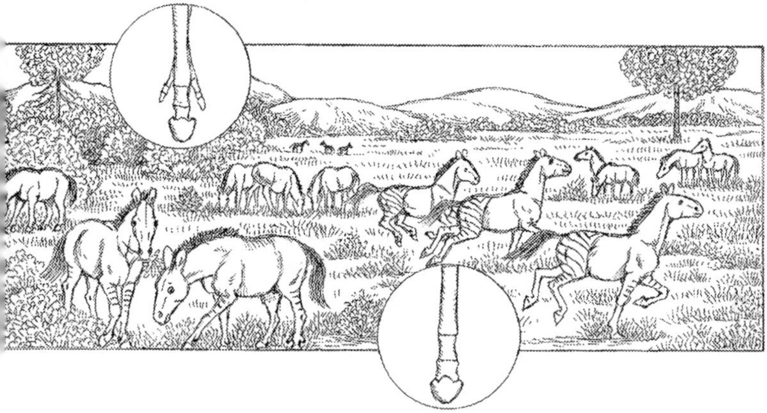

American mammals to extinction.* While it is commonly known that creatures like woolly mammoths and saber-toothed cats went extinct at this time, it's perhaps a bit more surprising to learn that modern horses also vanished completely in the New World and did not reappear until the Spanish conquistadores reintroduced them in the early 1500s.

Sadly, of the thirty-four recognized genera in the family Equidae, only one survives. What does remain, however, from this once diverse and widespread group, is a transitional fossil record that is unsurpassed in its ability to shed light on the relationship between environmental change and the accompanying structural

*Dr. MacPhee was troubled by the fact that these well-established and often formidable mammals could be wiped out in a geologic instant by Paleo-Indians wielding pointed sticks. "Why haven't similarly equipped Bushmen driven any large African mammals to extinction?" MacPhee asked during a symposium on Pleistocene extinctions, before offering an alternative hypothesis. "What if humans or the domestic animals they brought with them to this continent were carrying something that the American mammals' immune system couldn't handle?"

modifications that can accumulate in generations of creatures living in those changed environments.

Unfortunately, no such easy-to-interpret transition exists for vampire bats or many other organisms, for that matter. Compounding the fact that bat bones are extremely delicate, fossils from creatures that inhabited tropical regions are relatively rare. This is primarily because the remains of the newly dead in such environments are usually dismantled, eaten, and destroyed—with little chance of preservation in the fossil record. The vast majority of vertebrate fossils come from creatures that lived near shorelines—beaches, rivers, or even ponds. Here, rapid sediment deposition could give the dead at least a small chance at becoming fossilized.

Regrettably, this phenomenon, along with the fact that most fossilized creatures had hard parts like shells or bones, led some paleontologists to describe the fossil record as "biased." Not a bad description, really. But problems arose when deceptively named creation scientists intentionally took the term (and others) *completely* out of context in an effort to discredit the theory of evolution and insert their own faith-based beliefs.*

So how *do* scientists think vampire bats evolved? In cases like this one, where the fossil record isn't very helpful, researchers often rely on knowledge of what works for organisms living today—preferably those that are closely related to the ancient creatures in question. Prehistoric environments are also important since they provide information on the climate and surroundings in which the ancient critters existed. For the most part, this tech-

*Although creation scientists pass themselves off as real scientists, when pressed (usually when sworn under oath) they invariably admit that they're not. In order to join their organizations (e.g., The Institute for Creation Research), they must take a vow that what is written in the Bible is the *only* truth, scientific or otherwise. But hey, if you believe that the earth is six thousand years old, you shouldn't believe in evolution either.

nique has led to the following hypotheses on the origin of blood feeding in bats.

In one scenario, protovampires fed on blood-engorged ectoparasites like ticks that were feeding on large mammals. Seemingly, the ectoparasite hypothesis was founded upon the knowledge that roughly 70 percent of bats are insectivores (although ticks are certainly *not* insects), combined with purely anecdotal reports that vampire bats consume parasitic moths. During my graduate studies, I added a modification to this hypothesis by suggesting that if protovampires had in fact gotten their first blood meal by dining on ectoparasites, then blood feeding might actually have originated during mutual grooming behavior. Vampire bats are extremely social animals and studies have shown that they spend approximately 5 percent of their time grooming each other. During such behavior, protovampires may have obtained their first taste of blood from the very same tick and bed bug species that commonly parasitize modern vampire bats (and, indeed, most bats).

Bat biologist Brock Fenton suggested that the small size of ectoparasites, combined with the difficulty of locating them on another animal, made the ectoparasite hypothesis improbable. He was also troubled by the fact that ectoparasites have a worldwide distribution, yet vampire bats are restricted to three New World species. In other words, if ectoparasites were found pretty much everywhere, feeding on all sorts of vertebrates, then why weren't there more species of vampire bats in existence? I'll address this question momentarily.

Another hypothesis on the origin of vampire bats was proposed by Fenton, who suggested that blood feeding might have evolved from protovampire bats feeding on insects and their larvae present in and around wounds on large mammals. These wounds, some of which can be quite gruesome, are the result of aggressive social behavior, thorns, or unsuccessful predation.* However wounds are inflicted, they can quickly become beacons for swarms of insects (like screwworm flies) searching for a meal or a warm, moist place to lay their eggs. According to Fenton, insectivorous bats feeding at wound sites may have received additional nourishment from the blood and flesh of the wounded animal itself—and at some point, these protovampires would have switched to feeding solely on blood. Fenton strengthened his case by citing the feeding behavior of oxpeckers, a pair of African bird species (genus *Buphagus*) related to the omnipresent starlings (family Sturnidae). Oxpeckers glean ectoparasites like ticks off large mammals and they're also reported to feed at wound sites and festering sores. Similarly, certain finches *(Geospiza)* remove ticks from giant Galápagos tortoises, which elevate their massive bodies on fully extended limbs to allow the tiny birds total access to the blood-engorged pests.

In any event, my problem with the wound-feeding hypothesis is that it proposes that vampire bats evolved in the face of environmental pressures that would have seemingly acted *against* the development of such behavior. Not only would potential prey need to be wounded but it would also have to be of conspicuously large size and relatively immobile. Because vertebrate blood is basically made up of water and protein, vampire bats cannot store energy in the manner of non-blood-feeding mammals (as fat, for example). This requires vampire bats to consume approximately 50 percent of their body weight in blood each night. Failing to do so, they can starve to death within two or three days. Now that is an *extremely* tough way to make a living—and studies have shown that vampire bats (especially young adults) may fail to obtain a

*A. R. E. Lewis told Fenton that approximately 10 percent of the African buffalo *(Syncercus caffer)* he studied in Tanzania had scars from unsuccessful lion attacks.

blood meal one out of every three nights that they hunt. I estimate that this figure would be prohibitively higher if the prey were required to have existing open wounds. Just as important, there are no living bats (nor mammals, for that matter) that are reported to feed at nonlethal wound sites.

Ultimately, it's extremely difficult to imagine what would have driven protovampires to abandon an insect-eating lifestyle for one dependent on locating large wounded animals on a nightly basis. I can't envision the selective pressure that would have led to this behavioral transition. As we'll see a bit later, the wound-feeding hypothesis also flies in the face of modern vampire bat behavior (sorry about that) since these bats can forage only for short periods of time each night. Finally, echolocation (highly evolved in vampire bats and all of their relatives) would have been useless in differentiating wounded from unwounded prey.

In the frugivore hypothesis, well-developed incisors used to slice through thick fruit rinds would have evolved in fruit-eating protovampires into the bladelike teeth that characterize modern vampire bats. Those who proposed this alternative scenario never discussed how or why this transition from fruit to blood might have occurred and the hypothesis remains undeveloped.

Some critics rejected the frugivore hypothesis on the grounds that vampirism never evolved in the Old World fruit bats* even though they too are known to possess large upper incisors. This reasoning is similar to rejecting the wound-feeding hypothesis on the grounds that worldwide distribution of ectoparasites fails to explain why there isn't a worldwide distribution of vampire bat species. Both of these arguments fall short because they suggest that evolution is somehow completely predictable (i.e., "If vampires evolved from fruit eaters in the New World, they must also have

*These bats belong to the family Pteropodidae and are commonly known as flying foxes.

evolved from fruit eaters in the Old World"). In reality, the exact set of circumstances that led to the evolution of blood feeding in New World bats (things like habitats, prey, and predators) was *not* present for the Old World bats. And even if those circumstances had been present, there would be no guarantee that blood feeding would have evolved again. As Stephen J. Gould explained in his outstanding book *Wonderful Life,* if we could somehow rewind the tape of the earth's history and then allow it to replay, there would be no guarantee that evolutionary outcomes would turn out exactly the same. Gould's point was that contingency (i.e., chance occurrences) had a great deal to do with which organisms survived to evolve over historical time. If, for example, the climate shifts that led to a reduction of North American forests had never occurred (or differed in some slight way), then, quite possibly, the modern horse as we know it would never have evolved. Similarly, if a meteor had missed the earth some sixty-five million years ago instead of slamming into an area near the current Yucatán Peninsula—maybe small, bipedal ostrich dinosaurs (Ornithomimosaurs) would now be trashing the earth's resources, instead of humans. With regard to the evolution of vertebrate vampires, far more subtle changes might have produced Old World vampire bats, vampire birds, or even blood-feeding rodents. For whatever the reasons, though, in the conditions that actually existed, a single group of New World leaf-nosed bats underwent the evolutionary changes that would ultimately result in the only obligate mammalian sanguivores.

As an alternative to previous speculation on vampire bat origins, I proposed the arboreal-feeding hypothesis. Basically, this suggests that protovampires may have been foraging in much the same manner as several species of vampire bat relatives do today, that is, by feeding in the trees on small vertebrates like birds, bats, lizards, rodents, and marsupials.

In that regard, *Diaemus youngi* and the hairy-legged vampire bat, *Diphylla ecaudata,* both hunt in trees—feeding primarily on perching birds. There are, however, significant anatomical and behavioral differences between them that provide hints about the evolution of their feeding behavior. While a number of primitive anatomical features indicate that *Diphylla* originated as an arboreal blood feeder, evidence points to a recent return to the trees for *Diaemus,* where its ability to prey on birds would have reduced competition with the wildly successful, terrestrial hunter *Desmodus rotundus.**

While the fossil record for bats is scanty, it does indicate that there were carnivorous members of the Neotropical bat family Phyllostomidae present ten million years ago, right around the time vampire bats are thought to have evolved. There were also major climactic changes occurring in South America at this time,

*Briefly, whereas *Diaemus* has fairly robust hind limb bones and can motor around quite efficiently on level ground, *Diphylla* has the relatively fragile hind limb bones typically exhibited by nonquadrupedal bats.

with evidence suggesting that formerly vast tracts of forest became isolated islands (refugia) surrounded by grassland. Similar to the horse evolution story in North America, these forest refugia and their surroundings may have become perfect staging grounds for evolutionary change—this time among the phyllostomids.*

Evidence indicates that at least one phyllostomid alive at this time was a carnivore. Because of its size, *Notonycteris* may have stalked its prey through the branches before subduing it with bites, in much the same manner as its oversized modern counterpart, *Vampyrum spectrum*. It is very likely that *Notonycteris* and other ancient phyllostomid relatives would have encountered an increasingly diverse arboreal fauna, as marsupials like opossums, as well as primates, sloths, and larger forms of birds, took up residence in the trees during this time. Some of these new inhabitants would have been too large for carnivorous bats to stalk and kill using previously existing attack strategies. Over time, isolated populations of some carnivorous phyllostomid may have undergone a behavioral shift that allowed them to exploit these larger animals as a food source.† Maintaining a stealthy approach to their potential prey, these protovampires might have started biting larger species as they slept in the branches at night. Similar to Brock Fenton's wound-feeding hypothesis, these early protovampires may have supplemented their normal diets with flesh and blood, in this case from the bitten animal. The tendency for creatures sleeping in the trees would have been to remain quiet and stationary—even after a bat bite. Sudden relocations or frantic movements by the stricken creatures would have attracted other, even more dangerous, nocturnal predators. For the protovampire, natural selection would have

*Islands, whether surrounded by water or grass, are wonderful places to observe evolution in action. It's probably no coincidence that Charles Darwin and Alfred R. Wallace independently developed the concept of natural selection—the long-sought mechanism for evolutionary change—after working on island chains.

†There is no evidence that this protovampire was *Notonycteris*.

favored adaptations that maximized the nutritional payoff, while minimizing the danger and the likelihood that the prey would move off. In that regard, teeth that could inflict painless bites and salivary anticoagulants to keep the prey's blood flowing would have been key adaptations, as would an ability to move nimbly along and under the branches where their prey slept. Anatomical modifications that allowed for spiderlike terrestrial locomotion may have evolved as the ancestors of the common vampire bat, *Desmodus rotundus,* and the white-winged vampire bat, *Diaemus youngi,* moved down from the trees. Quite possibly these bats would have modified their arboreal blood-feeding techniques to exploit a new source of blood—ground-dwelling vertebrates like procyonids (raccoons and their cousins) or cow-sized, herbivorous tanks called glyptodonts (which are related to modern armadillos).

In nature, this type of coevolution between parasites and their hosts (or predators and their prey) is the rule rather than the

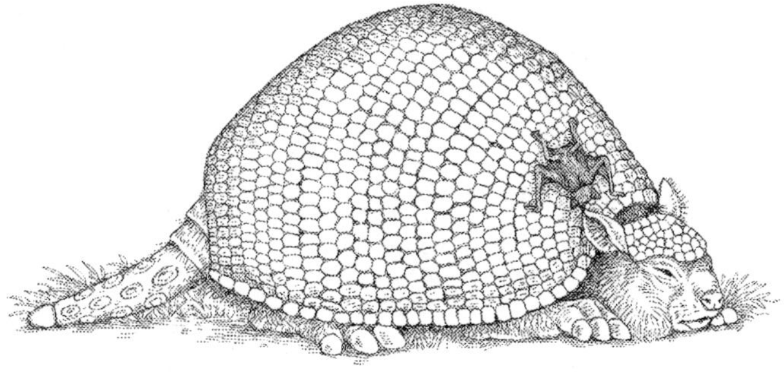

exception.* In this case, these early vampire bats were simply filling an open niche by exploiting a previously unexploited resource—vertebrate blood.

As far as what *really* happened, that's still open to debate. But given the opportunistic nature of modern vampire bats, it wouldn't be a stretch to learn that ancient protovampires actually exploited some combination of wounds, ectoparasites, *and* larger forms of arboreal fauna on their evolutionary road to becoming modern vampire bats. Perhaps, though, blood-feeding bats came about through a completely different scenario, as of yet unknown to scientists and leaving the question of vampire bat origins open to further debate and future research.

Some of you may have wondered why I chose to describe the various scenarios for the origin of blood feeding as, for example, the arboreal-omnivore *hypothesis* and not the arboreal-omnivore *theory*. Although you wouldn't know it from seemingly countless examples in the literature, there is a *major* difference between a hypothesis and a theory. A hypothesis is really a "best guess," based on an accumulation of evidence (generally observations or experimental data). Hypotheses are starting points as researchers attempt to answer questions that arise in science, such as "How did vampire bats evolve?" Often short lived, they're commonly modified as new evidence accumulates. Theories, on the other hand, can start out as hypotheses, but they are *far stronger*—having withstood

*According to some researchers, macroscopic blood feeders, like those discussed in this book, are more accurately defined as predators than as parasites. Dr. Stephen Spotte summed up the distinction between the two. "The modern definition of a parasite is an organism that is in intimate physiological contact with its host. The malarial parasite, *Plasmodium,* for example, living in the salivary glands of a mosquito and then able to camouflage itself once it gets into the human bloodstream—now that's a parasitic arrangement of a very high order. The mosquito, sucking blood from another animal for a minute or two, that's a blood predator." Ultimately, though, to keep from confusing the issue when using terms like *ectoparasite* or when quoting from interviews, both *parasite/host* and *predator/prey* will appear throughout this book, with the latter referring to specific instances where one party is killed during an initial transient encounter.

the test of time (and vigorous scientific challenge) and garnering support from numerous and varied fields of study. For example, a *theory* exists that life on this planet evolves. There were several *hypotheses* as to just how this scenario came about—with natural selection being the mechanism best supported by the evidence.

However vampire bats evolved, there is fossil evidence that at least three additional blood-feeding phyllostomids lived somewhere between two million years ago and six thousand years ago. Interestingly, this includes at least two North American species (*Desmodus archaeodaptes* and *Desmodus stocki*). In all likelihood, the extinction of these ancient vampires (which had ranges extending from California to Florida) was linked to climate changes, as a cycle of cooler summers and warmer winters transitioned into our current climate of hotter summers and cooler winters. Unable to find enough food during the winter or migrate long distances, their highly specialized blood-feeding diet would have sealed their fate, preventing them from packing on the fat necessary to survive a winter's hibernation.

The presence in the fossil record of a giant vampire bat, *Desmodus draculae,* suggests that this creature was feeding on megamammals like the giant ground sloths and heavily armored glyptodonts. *Desmodus draculae* was significantly larger than modern vampires and there is some evidence that they lived as far north as northern West Virginia.

Presumably, all of the ancient vampire bat species died out in North America following the great megafaunal extinctions of the late Pleistocene. One vampire bat expert, however, was convinced that at least some of them had not gone extinct.

"I think *Desmodus draculae* might still be alive," Arthur Greenhall told me during lunch one afternoon in Boston.

"How do you figure that?" I said, after nearly choking on my sandwich.

Greenhall explained that there were still regions in South

America where very few people visited—the deep Amazon, and parts of Brazil's Planalto Central, for example.

"Besides," he added, "*draculae* bones have been uncovered right alongside the remains of living species."

These facts, combined with the relatively recent discoveries of "living fossils" such as the coelacanth (a lobe-finned fish thought to be extinct for sixty million years) apparently gave the old vampire meister at least some hope that *Desmodus draculae* might still haunt the South American wilderness.*

"If only," I replied. If only.

*Living fossils aren't confined to the animal kingdom. The dawn redwood, *Metasequoia glyptostroboides,* was rediscovered in China in the 1940s.

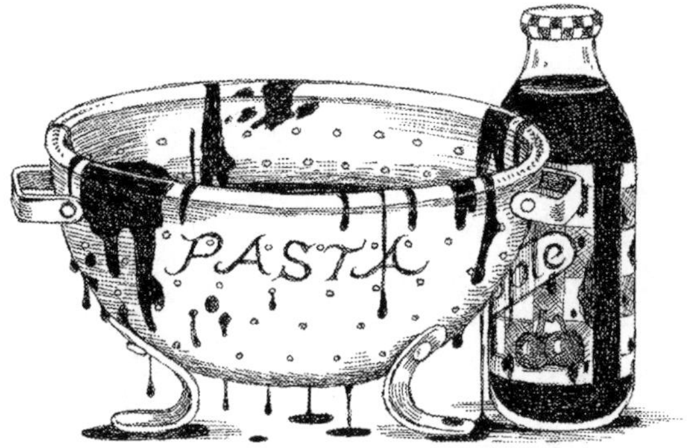

I have not seen anything pulled down so quick since I was on the Pampas and had a mare that I was fond of go to grass all in a night. One of those big bats they call vampires had got at her in the night, and, what with his gorge and the vein left open, there wasn't enough blood in her to let her stand up.

–Bram Stoker

3.

SNAPPLE, ANYONE?

It was still dark when we arrived at the slaughterhouse, but there were already several cars parked outside the nondescript, single-floor structure. We stood in the parking lot, finishing hot drinks—and not saying much—until the loud clank of a metal gate caused me to jump slightly.

"You need *more* caffeine," my companion said, sipping her decidedly low-test tea-water thing.

I responded by pouring the remainder of my cup onto the gravel.

As we approached the door, the aroma of my coffee gave way to the acrid tang of disinfectant and something else, something metallic, coppery.

There were new sounds coming from within the building, shouts and a deep organic vibrato.

At the abattoir, the workday was about to begin and we entered without knocking.

My companion that morning was Cornell undergrad Kim Brockmann, whom I'd met several months earlier when she began showing up at our weekly Zoology Journal Club. After one such meeting I had inquired if anyone might be interested in volunteering to help me maintain some vampire bats that I hoped to bring back from Trinidad. Kim's hand shot up without hesitation. Now, bundled up against the predawn chill, she was clutching six large plastic bottles and a spaghetti strainer, and I wondered if this is what she'd had in mind.

In Trinidad, obtaining blood on a daily basis hadn't been much of a problem, basically because Farouk Muradali, and more recently, his right-hand man, Keith Joseph, had been maintaining colonies of the common and white-winged vampires on and off for twenty-five years. During my first visit to Farouk's lab at the National Animal Disease Center, I was actually rather bowled over at their success in keeping the white-winged vampire, *Diaemus youngi*, alive in captivity. Several references I'd previously read (including one coauthored by my friend Arthur Greenhall) stated that these bats could not be kept in captivity for any length of time.

"Yes, we know all about those references," Farouk said with a dismissive wave of his hand. "They're one of the reasons why so little is known about *Diaemus*."

I bent over to examine a cluster of shapes that were gathered in the upper far corner of a spacious rectangular cage. All of the bats were asleep, except one, and he was watching me—mouth slightly open, sharp triangular teeth—strikingly white. *Desmodus rotundus* had black, beady eyes and an unmistakable air of intelligence. It was a look that Kim and I would become extremely familiar with over the next three years, one that always gave me the impression that the bats were waiting for me to make a mistake—the kind that would result in either their escape from captivity or the infliction of a savagely deep bite.

Farouk nodded toward the cage and continued. "*Desmodus* is not a picky eater. If you capture one tonight and put out cow blood for it tomorrow night—it will drink until it's full."

I nodded, recalling that these vampires had been given the species name *rotundus* because of their round-bellied appearance. Unfortunately, the naturalist Geoffroy never realized when he named them that their rotund abdomens were due to a gastrointestinal tract bloated with blood. Had he dissected a specimen he would have certainly noticed (as did Darwin's friend and supporter Thomas H. Huxley) that the common vampire's esophagus didn't empty directly into the stomach, a feature that typifies all mammals. Instead, the lower end of the esophagus ended in an inverted T, one serif leading to the stomach, the other leading to the intestine. Furthermore, the stomach wasn't J-shaped (as is seen in most mammals). It was tubular, a blind-ended U that was nearly two-thirds as long as the ropelike intestine that it closely resembled.

Not surprisingly, researchers who sought to determine the route of ingested blood in vampire bats found that the going was weird indeed. In an experiment using barium-laced cow blood, an X-ray machine, and five (presumably grumpy) common vampire bats, G. Clay Mitchell and James Tigner determined that freshly ingested blood moved from the vampire's mouth to the esophagus and then into the intestines *before* passing into the stomach.

We know now that these variations in digestive anatomy and physiology (like other anatomical and behavioral adaptations) are related to the vampire bat's unique lifestyle. In all other mammals, a primary function of the stomach is bulk storage of food—with some breakdown of that food (digestion). There is little transfer of nutrients or other material from the lumen of the stomach to the circulatory system (which then distributes it to the body).* This last, and generally overlooked digestive system function (absorption), is usually carried out by the small and large intestines and the network of blood vessels that supply and drain them. In vampire bats a key role of the stomach appears to be the rapid absorption of water (which makes up approximately 80 percent of the ingested blood volume). This excess water is carried to the kidneys (via the circulatory system once again) where much of it is converted into urine (i.e., water plus dissolved nitrogenous waste products). As mentioned earlier, since blood contains a negligible amount of fat, a substance in which energy is typically stored for later use, vampire bats are required to drink about half their body weight in blood each night.† This sudden weight gain is an extremely dangerous proposition for an animal that might be required to take flight at a moment's notice. Because of this, an important adaptation for blood-feeding bats is their ability to shed weight as quickly as possible, and their digestive and excretory systems reflect this emphasis. Researchers who have ob-

*A lumen is the space inside a tubular structure like the stomach, intestines, or blood vessels.

†According to a 1962 paper by legendary bat biologist William Wimsatt and his technician-colleague, Anthony Guerriere, a single vampire bat consumed 7.3 liters (fifteen pints) of blood per year, which worked out to about 25 gallons over a thirteen-year life span. George Goodwin and Art Greenhall took into consideration the fact that vampire bites continue to bleed long after the bat has finished feeding. They estimated the annual blood loss from each vampire bat to be 5.75 gallons—considerably more than the bat consumed.

served common vampire bat feeding sessions know that the bats begin urinating well before they've finished feeding. Along with modifications to the typical mammalian stomach and intestines, evolution has cranked up the vampire bats' excretory system, enhancing its ability to deal with its owners' rather unique dietary requirements.

Vampire bats get most of their nutrition from protein (in this case, hemoglobin and blood plasma proteins like albumin, fibrinogen, and globulin). These proteins are composed of smaller subunits called amino acids. In mammals, a problem arises when these nitrogen-bearing amino acids are broken down during digestion, releasing the toxic compound ammonia. Most mammalian excretory systems deal with ammonia by having the liver quickly convert it into a less toxic substance—urea. Urea is not only safer to have circulating around the body than ammonia but can also be more easily excreted by the body as urine (which is basically urea diluted with water extracted by the kidneys from circulating blood plasma).

As a vampire bat begins to feed, the kidneys immediately begin producing an extremely dilute form of urine in order to shed excess weight (approximately 25 percent of the blood volume consumed is excreted as urine within the first hour after feeding). Soon, though, the excretory system shifts into a very different gear. The urine produced becomes more and more concentrated as the kidneys work frantically to eliminate the rapidly accumulating urea without causing dehydration in the bat.

Because of the vampire bat's need to excrete massive amounts of urea, water loss is a real and constant problem. Dehydration may have been one of the key factors that limited vampire bat ranges to regions that had a high relative humidity and may be one more reason why prehistoric vampire bats disappeared in North America. In a similar vein (sorry, I had to get that out of my system), vampire bats are restricted to relatively short-duration flights each night.* Presumably, this limitation is due to the evaporative water loss that typifies bat flight. So, although many bat species are known to migrate or undertake lengthy nocturnal flights, a sanguivorous diet appears to have selected against these behaviors in vampire bats.

In 1969, Cornell University vampire bat expert William Wimsatt and his coauthor William McFarland put things in perspective: "In a very real sense the vampire bat can be considered to inhabit a desert in the midst of the tropics. But the desert is delimited not by environmental aridity, but rather by the nutrition and behavior of the vampire."

In both the vampire bats' digestive and excretory systems, there have been evolutionary trade-offs related to blood feeding, as speed of food processing has been increased at the expense of nutrient storage. This type of trade-off is a hallmark of biology. Some organisms adapt to the changing environments in which they exist (e.g., vampire bats are able to feed on blood in a humid environment but can't migrate or fly for extended periods). More often, though, organisms are unable to adapt—like many of the browsing mammals that died off as North American forests transitioned into grasslands. Although it's certainly more fun to think about the big, sexy mass extinction events (like the one that occurred around sixty-five million years ago), the vast majority of species that have ever lived on this planet apparently disappeared rather gradually. Most extinctions, it seems, were accompanied by a whimper, not a roar.

Farouk and I moved past the *Desmodus* cage to a smaller unit and peered in. Like their "common" cousins, all of the white-winged vampires were asleep except for one—but that was where the similarities ended. Where *Desmodus* has sharp, hard features

*This restriction would appear to select against the previously discussed wound-feeding hypothesis.

(I've always maintained that they *looked* the part of vampires), *Diaemus* reminds me of a stuffed-animal version of a vampire bat. Their face is much softer—the sharp angles seem smoothed out and the eyes are huge and gentle looking. Their demeanor reflects these anatomical differences. In three years of handling *Diaemus* in captivity, they never attempted to bite me. Not once.

Farouk shook his finger at me and then continued his lesson. "Now, if you leave out the same tray of cow blood for newly captured *Diaemus*, you'll have dead bats in two nights."

I soon learned that Farouk's "secret" for success was that each night he hand fed his newly captured white-winged vampires using a five-cubic-centimeter syringe full of cow blood. If the bats refused to eat, he didn't force the issue but instead provided them with the opportunity to feed on a live chicken. Over the course of several weeks, even the most finicky of his "babies" got the message and soon enough he had them feeding on cow blood that he'd poured into an ice-cube tray. He supplemented his captive bats' diet once a week (and probably on holidays) with feeding sessions on live chickens. These usually consisted of four hungry vampire bats to one soon-to-be-anemic chicken.

Now it was all starting to make sense—why 99 percent of everything that had ever been written about vampire bats dealt solely with the common vampire bat, and why even bat experts had told me that all three vampires would act similarly. *Desmodus rotundus* had been maintained successfully in captivity for nearly sixty years—with some individual specimens surviving for as long as twenty years. Additionally, these bats were numerous across their widespread range and therefore relatively easy to obtain (hence the name "common," I guess). They were also cheap to feed—as long as you had a ready supply of cow blood on hand. Plus, they were interesting as hell, with a slew of unique behavioral, anatomical, and physiological features.

The other blood-feeding bats, *Diphylla ecaudata,* the hairy-legged vampire (which didn't live in Trinidad), and its white-winged relative, *Diaemus youngi,* were far more rare (relatively speaking) within their limited ranges. They were much more difficult to locate and capture than *Desmodus,* and reports on the difficulty maintaining them in captivity only served to compound the problem.

As a result, most researchers (with a few notable Mexican and South American exceptions) simply avoided working on two of the three vampire bat species. Hence, there were relatively few studies on these bats—especially on topics like comparative anatomy or behavior. Thanks to Farouk Muradali, though, who had graciously decided to let me in on his secret for maintaining *Diaemus* in captivity, the door would soon be wide open for the comparative work I'd proposed to undertake.

Hand feed them until they start guzzling cow blood. How simple, I thought, until Farouk allowed me to do just that with one of the bats his crew had captured the night before. With no hesitation,

I showed off years of animal handling experience by mishandling the syringe and squirting the poor creature with an eyeful of cow blood.

"Must be the gloves," I said.

Farouk shot me a sideways glance, then smiled. "Yes, that must be it," he said.

Luckily for the bats, I got better.

"Slaughterhouse Bob" reminded me of Popeye with an extremely selective case of Tourette's syndrome. He was generally a friendly sort of fellow and he seemed genuinely amused that a couple of Cornell types showed up each week at 5 a.m. looking for cow blood. At the first sight of a health inspector, though, Bob's conversation would undergo a seamless transition into a machine-gun barrage of obscenity that would have made the most hardened dockworker blush like a ten-year-old girl. It was an uncomfortable moment for the health inspector as well since the cursing was clearly directed at him. Additionally, while Bob was ranting he was also wielding a nasty device known in the slaughterhouse trade as a captive bolt stunner. This was an instrument that looked like a cross between a power drill and Dirty Harry's .44 Magnum.*

Typically, Kim and I would stand back as Bob herded a single captive cow into the "stunning box," a heavy-duty, steel-railed

*In addition to the captive bolt stunner, slaughterhouse personnel also use tools like "brain suckers" and "bung ring expanders." The former has a rather self-explanatory function, while the Jarvis BRE-1 "mechanically seals the bung with a ring." According to the Jarvis Web site, this "reduces human error during bunging" (a big problem for most of us, especially after a few drinks). Not to be outdone, poultry slaughterhouse personnel wield their own line of rude-sounding gear. These folks systematically turn chickens into chicken parts with instruments like "picking finger cutters," "lung guns," and "vent cutters" (which also comes in the larger turkey model, a popular gift around Thanksgiving time).

enclosure, designed to keep the doomed animal from doing anything more than just standing there. This process generally began right after the health inspector realized that there was somewhere else he needed to be. Stepping up onto the bottom rail of the box, Bob placed the business end of the captive bolt stunner against the cow's skull, at the center of an imaginary X formed by the animal's eyes and the base of its horns. Bob never appeared to rush and he never "chased the cow's head," taking his time for one clean shot.

The concussive impact of the bolt stunner discharging sounded like a small-caliber pistol firing in an enclosed room (which in some sense is exactly what was happening). The results were as instantaneous as they were stereotypical. The animal collapsed, its brain penetrated by the steel bolt, which had already retracted back into the instrument.

Bob bent down, checking for eye reflexes by touching the cow's cornea. Anything resembling a blink would have meant that the creature hadn't been properly stunned—although in three years of visits to the slaughterhouse we had never seen this happen.* Having been assured that the enormous bovid wasn't about to right itself, Bob climbed agilely into the stunning box and disappeared behind the cow's hindquarters.

"This here's the most dangerous part of my job," came a muffled voice from somewhere just south of rump roast. "These cows still got nerves."

"That's for sure," I said, finally getting to use the anatomical knowledge I'd accumulated over a long collegiate career.

"One stray kick can break a man's back."

*Recently some slaughterhouses have moved away from penetrating captive bolt stunners, preferring nonpenetrating stunners because of concerns about bovine spongiform encephalopathy, or BSE (also known as mad cow disease). In *No Country for Old Men*, Cormac McCarthy's angel of death, Anton Chigurh, showed no such concerns, using an air-powered version of Bob's bolt stunner to dispatch his victims.

I pondered that image for a moment. "And that would suck," I added thoughtfully.

My colleague Kim (an aspiring anatomist herself) nodded in agreement. "Definitely."

In any event, I always got a bit antsy when Bob jumped in with a brain-bonked cow, and similarly, I always felt relief when Elsie rose from the floor, hind limbs first, cranking toward the ceiling under the power of a motorized block-and-tackle set.

Less than a minute later, the insensate animal had been hung so that its head was suspended above a large plastic barrel. Then, with one expert slice of his knife, Bob would sever one of the cow's jugular veins, stepping out of the way just in time to avoid the powerful torrent of blood that splashed into the blue container.

Once the cow had been fully exsanguinated (and just about the time that Bob started reaching for the "carcass-splitting saw"), Kim and I slid the sloshing barrel of hot blood to the opposite side of the room. Clad in fishing waders and rain gear, our hands were gloved in rubber for reasons that Mr. Playtex couldn't have imagined in his wildest nightmares. Even Bob shook his head in disgust—then he fired up the "Ronko Carcass Master 5000" and began the noisy process of carving Elsie into easy-to-carry pieces.

Standing over the barrel, Kim and I took turns using a metal spaghetti strainer to agitate the blood. By doing so we were actually speeding up the natural clotting process—which had been chemically triggered as soon as the blood left the confines of the severed vein. Although unable to stem the flow of blood from a traumatic wound to a major blood vessel, the hemostatic (clotting) mechanism we were currently stimulating did an extremely efficient job of preventing excess blood loss after minor injuries. For example, a divot-shaped wound of the size inflicted by vampire bats (approximately three millimeters in diameter) would be expected to stop bleeding within one or two minutes. This is not the case, however, in instances in which the wound is created by one of nature's blood-feeding specialists (e.g., leeches and vampire bats). Evolution has provided these creatures with a number of ingredients in their saliva that can interrupt the process of blood clotting for up to several hours. The end result is that the blood feeder is able to drink its fill, having temporarily halted the very same clotting process Kim and I were currently accelerating with our colander.

Vampire bats, feeding at a bite they've inflicted, use their tongues to draw out the blood. Contrary to popular belief, they do not suck blood from their victims. In fact, the physics involved is very similar to what happens when a phlebotomist draws up a patient's blood into a capillary tube. Basically, these thin glass tubes work because their inner diameter is so small that the force of attraction between the blood and the glass is greater than the downward pull of gravity. Thus, the blood pulls itself up the inside of the tube, filling it to a considerable degree.

In the case of vampire bats, the pistonlike motion of the bat's tongue causes blood to flow (via capillary action) along a pair of grooves located on the bottom of the tongue and directly into the bat's mouth. There's even a cleft on the bat's lower lip and a space between its lower incisors to facilitate the blood flow. While feeding in this manner, saliva is constantly applied to the wound.

Vampire bat saliva contains several ingredients that inhibit the body's normal clotting mechanisms. One such anticoagulant compound works by preventing blood platelets from clumping together—an important step in the formation of a plug that will eventually become a blood clot. Meanwhile, another salivary ingredient inhibits the torn blood vessels from constricting—a process that normally reduces blood flow to the wound site and thus from the wound itself. Finally, an enzyme that medical researchers would christen desmokinase (and, later, desmoteplase or DSPA for *Desmodus rotundus* salivary plasminogen activator)

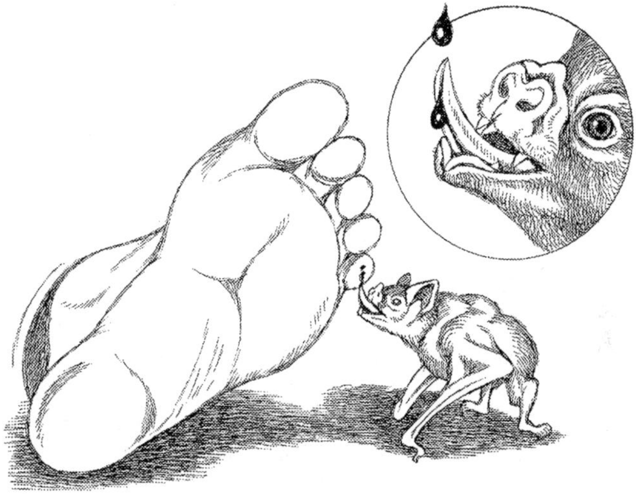

breaks down the protein framework upon which the remainder of the blood clot forms.

Primarily because of its antihemostatic properties, vampire bat saliva has drawn considerable attention from the medical community as a potential treatment for certain strokes, namely, those in which a blood clot inhibits blood flow within the brain's blood vessels. In these instances, cells in the region of the brain on the downstream side of the clot are denied oxygen and nutrients. If this blockade continues for long enough, the cells die and the function they were responsible for is impaired. Traditionally, stroke victims have been treated with a compound called tissue plasminogen activator (t-PA). Unfortunately, t-PA must be administered within three hours of the stroke to be effective. After that, the risk of bleeding into the brain increases, and as a result so does brain cell death. Since the average stroke patient waits more than twelve hours before going to the emergency room, t-PA is rarely administered and cannot be considered an effective treatment for the nation's third-largest killer (after heart disease and cancer). Unlike t-PA, though, studies have shown that the vampire bat–derived DSPA (an extremely potent clot buster) can be administered up to nine hours after a stroke has occurred and has no detrimental effects on brain cells.

Since vampire bats often lick the site *before* inflicting their bite, there has also been some speculation that their saliva might contain either a pain-killing agent that prevents their prey from feeling the bite, or an enzyme that could function to soften up the potential bite site. Even without a painkiller or a skin-softening enzyme, the vampire bat's razor-sharp teeth are likely capable of producing a wound that causes little or no pain to the prey.

In what amounts to a strange sort of payback, one modern technique used for vampire bat eradication involves the use of the anticoagulant warfarin. Isolated from a clover mold, warfarin has been marketed for humans since the 1950s as Coumadin. It remains (along with the lung- and intestine-derived compound heparin) a popular blood thinner for millions of patients prone to strokes or blood clots.

Vampire bats treated with warfarin are subject to no such medical benefits. After capturing the bats in mist nets, the animals are painted with a mixture of warfarin and Vaseline and then released so that they can fly back to their roosts. Since vampire bats spend a considerable amount of time grooming one another, the toxic paste is soon spread throughout the members of the roost—with deadly results. Bats soon perish as the ingested warfarin induces massive internal hemorrhaging, causing them to bleed to death.

Although some might consider vampire bats dying from a clot-busting anticoagulant to be poetic justice, others might consider it to be a cruel waste of neat bats. In any event, warfarin paste is certainly several steps up from the days when cries for vampire bat heads sent folks scurrying for dynamite, poison gas, and flamethrowers. As long as vampire bat control personnel are applying the paste to the correct bats, this eradication method is relatively species specific. The drawback is that it's successful only in places (like Trinidad) where individuals are trained to capture the right bat species, untangle them from mist nets (no easy task), and then apply the poisonous paste. All of this must be accomplished without injuring the bat, getting hammered by its powerful jaws, or painting a nonvampire bat by mistake.

A related, but less-cost-efficient, method of vampire control involves the inoculation of livestock with low doses of anticoagulants. Vampires feeding on the inoculated bovine blood suffer the same hemorrhagic fates as those grooming warfarin paste off their roost mates. While this systemic method eliminates the need to capture and correctly identify vampire bats, it does require the treatment of the entire herd in order to be effective.

In the end, both of these methods are successful because of the low reproductive rate in vampire bats. Like the vast majority of bats, vampires give birth to only one pup per year—a far cry from other mammal pests like rodents, who can crank out babies faster than a baseball player can split sunflower seeds.

Back in the slaughterhouse, Kim and I used the same colander to strain off the strange, woven clots that materialized in the barrel—squeezing out the blood they held, before discarding the sponge-like clumps in a waste bucket. After about fifteen minutes of this fun, the defibrinated blood that remained in the barrel (i.e., the blood minus the clotting factors and proteins making up the clots) was poured, still warm, into the two-gallon plastic containers we'd brought with us. By stimulating clots to form, and then removing them, we were assured that our defibrinated blood would remain liquefied and clot free during storage, and when we placed it out to feed our bats.

Although we didn't realize it at the time, a similar method was employed between the 1820s and 1920s to defibrinate a human donor's blood prior to transfusion. In the days before the medicinal use of anticoagulants, donated blood was collected in a bowl, whisked, and filtered before being transfused into a recipient.

Some researchers use an alternative method to facilitate the storage of blood (for vampire bat meals and other purposes). The technique involves "citrating" the blood by adding the compound trisodium citrate to it. This also prevents the formation of clots, and although we never employed this method, in hindsight it could have provided our captive vampires with a slightly more nutritious meal. This is because, unlike our whisk and filter method, the clotting proteins aren't actually removed from the citrated blood.

After returning to our lab at Cornell's College of Veterinary Medicine, Kim and I transferred the blood into several dozen Snapple bottles we'd collected earlier from the cafeteria (yes, we cleaned them first). We froze the blood-filled bottles, thawing one out each morning so that the liquid would reach room temperature by nightfall. That was when we fed our vampire bat colonies—pouring the blood into an ice-cube tray and elevating it with a wooden block so that the roosting bats wouldn't have to strain themselves while they ate. As Farouk had done in Trinidad, we supplemented the diet of our white-winged vampires with a live chicken (once per week and on holidays). This turned out to be a vital step in maintaining our vampire colony, as I found out three years later.

Shortly after passing the bats off to another Cornell grad student (who had proposed a study on their digestive physiology), I received a rather frantic call from Kim. I discovered that the new researcher had not only relieved my friend of her bat-keeping duties but had decided to suspend the colony's live chicken supplement (basically to save a few bucks each week while eliminating the far from insubstantial hassle of dealing with live chickens). Within ten days, vampire bats began dying at an alarming rate—a trend that stopped immediately after a "talk" with the grad student led to a resumption of weekly chicken dinners for the colony.

During the three years that we maintained our colonies of common and white-winged vampire bats, it's safe to say that we saw some strange stuff, much of it relating to feeding behavior or social interactions between roost-mates. We found out later that Farouk and his Trinidadian bat crew had already noted much of what we were observing at Cornell. Their reluctance to publish, however, made it news to us, and we were grateful that these bat experts had (for some reason) decided to take us on as collaborators and coauthors. There were numerous occasions when something very much like the following exchange took place over a crackling long-distance phone line.

"Yes?" Farouk's Trinidadian accent made it sound more like *yes-ah.*

"Farouk?"

"Yes."

"You're not going to believe what we just saw."

Silence.

"I think *Diaemus* is mimicking chicks. They're snuggling right up to these hens—then biting them on the chest. It's unfrigginbelievable!"

Silence.

"Farouk?"

"Yes."

"Have you ever seen that before?"

"Yes. The bites are on the brood patch."

"Oh . . . Cool. Okay, I'll talk to you soon."

"Yes." Click.

I've always considered my friend Farouk Muradali to be one of the most generous and nurturing people I've ever met. But to say that he is a man of few words . . . well, you get the picture.

My collaborators and I also learned from the start that Arthur Greenhall had been right about the significant differences that existed between vampire bat species (in our case, between *Desmodus rotundus* and *Diaemus youngi*)—and we would discover that most of this variation was related to the bat's preference for either mammalian or avian blood, respectively.

"*Diaemus* doesn't jump," Farouk had said (in what would become his equivalent of the Gettysburg Address). And after a hundred-plus trials on our miniature force platform, we had to agree. But why was this so?

Initially, we tested our system out with the common vampire bat, *Desmodus,* and as in previous studies, we confirmed that these bats could make spectacular, acrobatic jumps, in any direction. Pushing off the ground with their powerful pectoral muscles,

Desmodus used its elongated thumbs (the last things to leave the ground) to impart precise direction to jumps that could reach three feet in height.

These amazing jumps, along with their ability to run at speeds of up to two meters per second, were adaptations for terrestrial blood feeding. They enabled the common vampire bat to escape predators, avoid being crushed by their relatively enormous prey, and initiate flight after a blood meal. The ability to feed efficiently on large quadrupeds is the primary reason why *Desmodus rotundus* has been so successful in terms of numbers and range, but in all likelihood this success was a rather recent development.

Until about five hundred years ago, *Desmodus rotundus* may have been anything but "common." In fact, populations would have been severely restricted not only by climate but by the finite number of large mammals that were present in any given area. Quite possibly the vampires would have been compelled (as they sometimes are today) to feed on smaller mammals as well as birds and other vertebrates like snakes and lizards.

Starting in the early 1500s, however, the influx of Europeans and their domestic animals into the Neotropics would have spelled big changes for *Desmodus,* as well as the other two vampire genera. Suddenly, enormous four-footed feeding stations would have sprung up in places where the pickings might have previously been sparse for thousands of years. Additionally, not only would there have been plenty of new animals to prey upon, many of these quadrupedal blood bags would have been penned in, making them super easy to find and ultimately making meal time a whole lot more predictable than it had ever been before. Populations of the opportunistic *Desmodus* would have exploded as more and more land was cleared for cattle farming. The more cows, pigs, and horses, the larger the vampire bat populations that could be sustained by their blood. Human victims weren't necessarily preferred, but they did give the vampires additional opportunities to feed, long before windows, screens, and protective netting would keep them at bay.

From the standpoint of the newly arrived humans, it must have seemed like yet another plague had descended upon them, for with the mysterious nocturnal attacks and the gruesome postbite cleanup came diseases, rabies being the most feared. Soon, stories of vampire bats, their gory attacks, and the horrible diseases that they inflicted, were making the rounds throughout Europe, and from there they spread to the rest of the world. What little scientific knowledge there was on the topic became blurred by misconception and misidentification, turning tales of these creatures into an unreliable blend of vampire fact merged with vampyre fiction.

Unlike *Desmodus, Diaemus youngi* (which resembles a winged teddy bear) has contributed little if anything to vampyre folklore. Perhaps they were once fast and aggressive, and maybe they even initiated flight similarly to their spring-loaded cousins. But now their movements are more deliberately paced and show little sense of urgency. When placed on the surface of our force platform, white-winged vampires would give a little hop or two, then scuttle off to find a dark corner in which to hide.

Watching *Diaemus* feed arboreally, we saw why they didn't need to catapult themselves into the air. Approaching a roosting bird from below the branch, white-winged vampires moved slowly and stealthily—advancing one limb at a time—and always keeping the branch between itself and the underside of its intended prey. Once situated beneath the feathered lunch wagon, *Diaemus* picked a potential bite site, usually on the bird's backward-pointing big toe (i.e., the hallux). This made perfect sense, since feeding from this particular digit kept the bat better hidden from above than if it had chosen to feed on one of the three forward-facing toes. After licking the chosen site for several minutes, an apparently painless bite was inflicted using the razor-sharp teeth that characterize all three vampire bat species. The bite was *never* violent and very often occurred as the bird shifted position slightly on its perch, as if reacting to some slightly uncomfortable irritant. Still hanging below its completely oblivious prey, *Diaemus* began feeding, and within five minutes it began peeing. It did so by extending one hind limb sideways and downward, deftly avoiding the embarrassment of soiling itself while eating. After feeding for between fifteen and twenty minutes, the bat would release its thumbs from a branch, hang briefly by its hind limbs, then drop into flight. Initiating flight in this manner, there was absolutely no need for *Diaemus* to jump, and so it didn't, at least not into flight.

On numerous occasions, we did observe *Diaemus* feeding on birds from the ground. Supporting its body in a low crouch (as compared with the extreme upright stance of *Desmodus*), the white-winged vampire was quite adept at hopping around (rather comically) in pursuit of a feathered blood meal. This behavior had not been reported in the wild and we used it to propose that the white-winged vampire bat had made a relatively recent return to the trees, thus avoiding competition with its ground-feeding cousin, *Desmodus*.

During these terrestrial feeding bouts we occasionally recorded behavior that approached chick mimicry on the "weird-o-meter." This occurred after the bat leaped or climbed onto the chicken's back, then scuttled forward, intent on biting the back of the bird's head or its fleshy comb. Male chickens mounted in this fashion quickly grew agitated and dislodged the bat with a shake and a peck. Hens, however, had an *entirely* different response. Rather than showing annoyance, female chickens quickly assumed a crouching posture that they maintained until after the vampire bat had finished feeding and hopped off. With a little research into poultry behavior, we learned that this was the identical posture

taken by a hen while being mounted by a male bird—for a completely different reason.

Another way that *Diaemus* differs from *Desmodus* and *Diphylla* is by the presence of a pair of cup-shaped oral glands located at the rear of the mouth. When *Diaemus* gets upset (or, as we observed, during dominance hierarchy behavior), the glands are projected forward and they can be seen quite easily when the bat opens its mouth. As it does, *Diaemus* produces a strange hissing vocalization that is accompanied by the emission of a fine spray of musky-smelling liquid from the oral glands. Although a detailed study remains to be performed, the oral glands of *Diaemus* appear to function in self-defense (like the scent glands of skunks) and as a means of communicating information like status, mood, and territorial boundaries to others of its kind.

Besides their actual ability to feed on blood, perhaps the most fascinating of all vampire bat adaptations is one that we observed only once in our colony of *Diaemus*.

In 1984, zoologist Gerry Wilkinson reported that vampire bats in the wild commonly share food by regurgitating blood. Wilkinson, who made his initial observations on *Desmodus rotundus,* determined that about 75 percent of the time blood sharing occurred between a mother and her dependent offspring (until about the age of one). In other instances, sharing took place between related or unrelated bats.

Gerry's results indicated that there were several reasons why this behavior occurred. Blood sharing between mothers and newborn pups presumably transfers nutrients and bacteria to the infant's digestive tract. In humans there are normally over two hundred species of bacteria living somewhere on or in our bodies (it's rumored that in some college dorms this number can hit five million species). In any event, these essential microbes (termed *bacterial flora*) are vital components of several physiological processes, most notably digestion.

In that regard, the mammalian small and large intestines (the

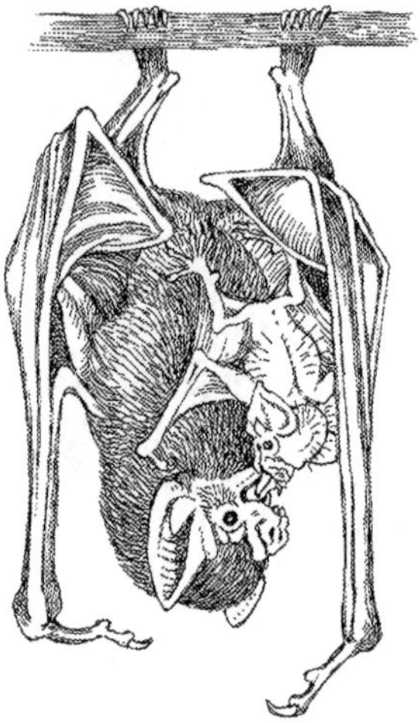

terms refer to diameter, not length) are home to billions of bacteria that have evolved a number of mutualistic relationships with their warm-blooded hosts. Often referred to as endosymbionts, these bacteria get food and a warm, moist environment in which to live. The mammals reap a number of benefits from the relationship, including the absorption of vitamins B_{12} and K, which are secreted by the bacteria as part of their day-to-day functioning.* Additionally, indigenous bacterial flora inhibit or kill nonindigenous forms, and they also prevent infection by stimulating the immune system to produce antibodies that can cross-react with potentially harmful nonindigenous bacteria, should they appear. In hooved mammals (i.e., ungulates), as well as wood munchers like termites, the presence of certain endosymbiotic bacteria en-

*In mammals, vitamin K is essential for the process of blood clotting, while a deficiency in vitamin B_{12} impairs red blood cell formation.

ables their digestive tracts to break down cellulose—the structural protein that makes up the plant cell wall. These bacteria are the prime reason herbivorous creatures are able to digest plant structures like leaves, stems, and wood. Since we don't have these specific endosymbionts, it's also the main reason why this type of "fiber" goes through humans like the vegan version of Roto-Rooter. Young herbivores aren't born with their bacterial flora either but instead obtain them from adults (like their mothers) through regurgitation or by consuming their feces (coprophagia). For this reason, termite "babies" denied their fecal formula are unable to digest wood and quickly starve to death.

Other related studies, by researchers such as Long Island University geneticist Ted Brummel, have shown that symbiotic bacteria increase the life span of fruit flies, even though the bacteria are apparently not involved in the digestion of plant matter.

Blood sharing between related and unrelated vampire bats also occurs on a reciprocal basis; that is, bats that were experimentally starved for one night before receiving blood from another nonrelated individual were more likely to donate blood to that individual when it was starved. This behavior is almost certainly related to the fact that the bats need to acquire a blood meal every night (and will starve to death in two or three days if they don't obtain one). So, over the course of their long lives (up to twenty years), there will presumably be numerous opportunities to receive and share food. The implication here is that *Desmodus* can remember past donors and can also recognize cheaters—those individuals who try to beat the system by rarely sharing blood. It's also interesting to note that although adult males share blood with females and young bats, they do not share with other adult males—which makes perfect sense. Why share food with someone you may be competing with for a mate?

There is evidence that both *Diphylla ecaudata* and *Diaemus* *youngi* also share blood (as I mentioned, we saw this behavior once in two captive specimens of *Diaemus*).* Unlike Wilkinson's in-depth study of *Desmodus,* however, this behavior in *Diphylla* and *Diaemus* has yet to be studied in detail.

This brings up an important point regarding original research—and one that I found quite helpful when I was just starting out in the field. I often advise students who are looking for research projects to seek out classic studies (like Gerry Wilkinson's) and then think about applying similar techniques to other organisms that have yet to be studied. Likewise, if the original research was done years earlier, new studies on the topic may warrant publication if the new researcher employs technology or methods that weren't around in the past (or asks questions that wouldn't have been asked in "the old days").†

Before leaving vampire bats to their bloody business, it's only fair that I mention the third genus, *Diphylla ecaudata,* the hairy-legged vampire bat. So named for the frill of hair that borders the back margin of its hind legs, *Diphylla* is thought to exhibit the most prim-

*This is what's known in the trade as an anecdotal observation, and clearly specified as such, it allows scientists to report information (usually among themselves) without submitting the material to the peer-review process. The understanding among researchers (although unfortunately *not* with some media types) is that anecdotal observations (and even pilot studies) are *expected* to be met with skepticism.

†For example, many anatomical papers published in the nineteenth and early twentieth centuries were purely descriptive in nature. They were full of exquisite illustrations (many of them hand-colored), but the accompanying text was often straightforward and generally interesting only to other anatomists. Nowadays, anatomy done for the sake of description is exceedingly rare. Far more frequently, researchers study the form of an organism (or its parts) as a way to propose and then answer an array of questions on topics ranging from evolution and ecology to biomechanics, paleontology, and behavior.

itive anatomical characteristics for its group.* In other words, scientists believe that *Diphylla* has undergone the least amount of evolutionary change from ancestral vampire bats—whatever they were.

One such primitive characteristic is that most bats (including *Diphylla*) have extremely thin hind limb bones (i.e., the femur, tibia, and fibula), and by thin, I mean that their diameter is quite small compared to their length. Scientists believe that this is an evolutionary trade-off related to flight. By having thinner, lightweight limb bones, bats have reduced their weight—an important factor for any flier. The downside of the trade-off becomes apparent if you watch

*The word *primitive* should not be used to describe whole organisms but rather only the specific characteristics of that organism not thought to have undergone relatively recent evolutionary change. For example, five digits is a primitive characteristic in humans since all primates share that trait (i.e., it hasn't evolved since the first primates). On the other hand (literally), a single-digit limb (like that found in horses) is considered to be a *derived* trait since it has undergone considerable evolutionary change from the multidigit condition seen in protohorses.

a bat moving around on the ground (which is something you generally don't see very often). In this regard, most of the eleven hundred species of bats can do little more than a clumsy shuffle when grounded, and even those that can walk are anything but graceful. Engineering models have shown that most bat hind limb bones did not evolve to withstand the compressive loads associated with walking. To demonstrate this for yourself, take a two-inch length of uncooked spaghetti and hold the ends between your thumb and index finger. Then bring your fingers together. You've just applied a compressive load to a model of a bat hind limb bone. Neat, huh? Now go pick up those pieces of spaghetti before someone steps on them.

As you have already learned, an inability to move about terrestrially is *not* a problem for *Desmodus* and *Diaemus*. These blood feeders are quite adept (and, in the case of *Desmodus,* even spectacular) as they walk, run, and hop about on the ground.

If you examine the hind limb bones of these two bats, it's not surprising that compared with *Diphylla,* they're thicker in *Diaemus* (i.e., they have greater diameter to length ratios) and *much* thicker in *Desmodus*—where they more closely resemble those of a small terrestrial mammal than they do typical bat hind limb bones. Apparently, stronger limb bones evolved in some vampire bats as they became adapted for current feeding strategies, namely, preying on large quadrupeds like pigs and cows. The evidence that *Diaemus* was once a terrestrial hunter lies in their robust limb bones, which seem overdesigned for their current arboreal roles. Additionally, *Diaemus* can scoot along quite well on the ground when it needs to, and it is quite capable of feeding while doing so.

The fragile hind limb bones of *Diphylla,* on the other hand, are clues to this bat's arboreal feeding habits; in other words, form reflects function. Unlike the requirements for walking and hopping, you don't need thick limb bones to hang under a branch when you feed since bones like the tibia and femur would be loaded under tension rather than compression. Researchers in the 1970s cited engineering models to hypothesize that hanging behavior in bats actually evolved because of thin hind limb bones, and you can demonstrate this concept with another short piece of pasta. Using the thumb and index finger (of both hands this time), gently pull on the ends of your two-inch length of experimental noodle. Unless you've twisted or bent the pasta by accident, you should be holding a piece of fracture-free fettuccine.* There—you've just modeled the tensile forces encountered by the hind limb bones of a hanging bat.

Besides *Diphylla*'s fragile hind limb bones, the hairy-legged vampire has another anatomical characteristic not seen in its blood-feeding cousins. In fact, this feature is completely unique to all other animals.

Many bats have a structure called a calcar, which is a bony or cartilaginous extension of their heel bone (the calcaneus). Since bat hind limbs are rotated up to 180 degrees from the typical

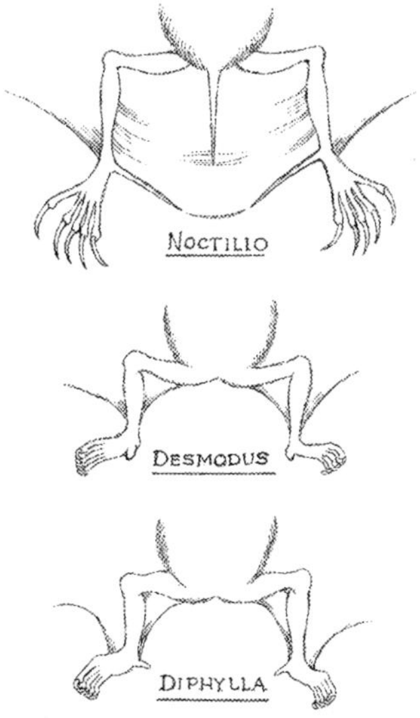

*Ronzoni No. 9 is actually perfect for this test.

mammalian condition (picture your knees facing backward), the calcar generally points toward the midline of the body. Its function is to strengthen and straighten the trailing edge of the tail membrane (uropatagium) that spans the space between a bat's hind limbs. Basically, the calcar increases aerodynamic efficiency by preventing this extra lift surface from flapping around during flight.

As one would expect, the calcar varies in size and shape among the eleven hundred bat species. In *Noctilio,* the fishing bat, for example, the calcar is a huge, bladelike affair. *Noctilio* uses it to get its substantial tail membrane out of the way while the gafflike hind limb claws skim the water's surface for prey.

It's also no surprise that the calcar is absent in bats that do not have a tail membrane. At least that's what I thought until I started examining preserved specimens of *Diphylla* at the American Museum of Natural History where I was working as a postdoctoral research fellow.

After determining that differences in performance existed between *Desmodus* and *Diaemus* ("*Diaemus* doesn't jump!"), I had started looking to see if these behavioral differences might be reflected by variations in their anatomy. While comparing the hind limbs of the three vampires, I noticed that the calcar was absent in *Diaemus* and reduced to a flaplike tab in *Desmodus*. Like I said, no big deal, when you consider that all three vampires lacked a functional tail membrane.*

The calcar of *Diphylla* was a completely different story. Not only was it present but it also stood out like a tiny finger. I immediately pulled out several additional specimens—just to make sure I wasn't just looking at one extremely weird individual. But in each

*Perhaps the lack of a tail membrane is yet another trade-off, this time between flight efficiency and quadrupedal locomotion. It's easy to imagine how the presence of an expanse of skin between the hind limbs might hamper a vampire bat's movement on the ground or among the branches.

instance, I saw the same digitiform structure. Next, I hit the literature, looking for any mention of *Diphylla's* calcar. "Small but well developed" ran the typical description—but nothing more. Finally, I called the vampire bat expert, Scott Altenbach, recalling that he had once maintained a colony of *Diaphylla ecaudata* in New Mexico. Scott had done the original work on quadrupedal locomotion in the common vampire bat in the 1970s and he'd joined us in Ithaca during our force platform project in 1993.

I remember a conversation something like what follows taking place over another crackling long distance phone line.

"Hey, Scott, did you ever take any photos of *Diphylla* climbing around the branches?"

"Yeah, but they weren't branches. We used wooden dowels."

"Well, check it out and let me know if your bats were using their calcars to grip the dowels."

"What?"

"I think *Diphylla* is using its calcar as a sixth digit during arboreal locomotion."

(Long pause)

"Scott?"

"I'll go get the photos."

(Sound of phone dropping)

Basically, what I'd proposed was similar to the story of the panda's thumb (popularized in an essay by Stephen J. Gould). The giant panda *(Ailuropoda)* feeds on bamboo leaves that it strips off branches, apparently with the aid of its opposable thumb (something not found in any other carnivores). Anatomists who examined the panda, however, found that things weren't quite as they seemed. The panda's thumb was actually a wrist bone (the radial sesemoid) that had become greatly enlarged. This allowed it to take on a new function—stripping off bamboo leaves.

Gould cited the panda's "thumb" as a beautiful example of how evolution doesn't create; it tinkers with what's already there (in this case, the panda's radial sesemoid bone), modifying it for a new function rather than creating a new structure from scratch.

Ailuropoda's odd little digit also presented some rather daunting questions for those who support a creationist view of how we got here. Basically, if there *is* an intelligent designer, why did he (or she) give the panda a jury-rigged structure for stripping leaves off branches? Why not just give the *Ailuropoda* a real thumb?

Back in Ithaca (and several minutes later), I heard scrambling and the sound of the receiver being picked up. "You nailed it, Bill—I've got some great shots."

"Excellent," I replied. "Do me a favor and send them to me at once."

Altenbach's black-and-white photos clearly showed a climbing *Diphylla* with its calcar tightly wrapped around a wooden dowel. I immediately put together a proposal to record this behavior in

the field, setting my sights on a visit to central Brazil. Since *Diphylla* didn't live in Trinidad, I contacted Brazilian researcher Wilson Uieda, who had been studying the hairy-legged vampire for years with his colleague, Ivan Sazima.

Outside the capital city of Brasília, at a ranch where the cattle were commonly plagued with *Desmodus* bites, Wilson and I set up my infrared video camera at sunset. We certainly weren't interested in the cows or even in *Desmodus* for that matter. Instead, we aimed our camera upward, into the branches of a fig tree, for it was there that the resident guinea fowl went to roost at dusk.

Several hours after nightfall, as I stared bleary-eyed at the camera's viewfinder, a pair of dark shapes flew past the sleeping birds.

"Wilson, check this out," I whispered.

My friend, who had been dozing on the chair next to mine, was instantly alert.

Less than a minute later, the aerial recon was performed for a second time.

Wilson whispered a single word. *"Diphylla."*

After that we saw nothing for several minutes—until a tiny pair of glowing spots appeared beneath one of the roosting birds. I hit the zoom on the camera, focusing in on the twin points of reflected light.

They were eyes!

Wilson traced a dark silhouette on the screen and I could just make out *Diphylla*'s upside-down head peeking out from the guinea fowl's feathery breast.

"Dinnertime," he said.

"This is different from *Diaemus,"* I responded.

Wilson replied with a smile.

Rather than feeding from below the branch, *Diphylla* was actually hanging from the bird—and photographs taken by Wilson Uieda and his colleagues at another site clearly indicated that *Diphylla* was using its opposable calcar to get a grip on the body

of its avian prey. Unlike the white-winged vampire, which generally fed from bites it inflicted on the toes of perching birds, many of *Diphylla*'s bites were made around the cloaca (the common opening for the digestive, urinary, and genital tracts found in many nonmammalian vertebrates, like birds).

Several days later, we visited a cave that was home to a small colony of *Diphylla*. Using the infrared camera again, we recorded three hairy-legged vampires as they moved across the stony ceiling. Not only were the bats walking upside down, they were moving backward (not really strange since bat knees face backward). What was unique was the way that they led with their hind limbs—carefully seeking a secure purchase before taking a step—and using their "sixth digits" like a rock climber would use his thumbs. After scrambling around the cave ceiling for a few minutes, the vampire bats tired of our intrusion and disappeared into a narrow crevice.

I left the cave elated that we'd been able to support my hypothesis with observations in the field. What had begun as a surprising observation back in New York City ended with the discovery that just like the panda's radial sesamoid bone, the hairy-legged vampire bat's calcar had been co-opted for a new role—as an opposable digit.

Even more important, although local scientists in places like Trinidad and Brazil had been aware of it for years, it wasn't until the very end of the twentieth century that the mainstream scientific community began looking at each of the three vampire bats as separate and quite unique. Thanks to researchers like Farouk Muradali, Wilson Uieda, and the late Arthur Greenhall, vampire bats are currently being studied with an eye toward variation rather than presumed similarity. By avoiding our tendency to lump things together, these scientists have increased our knowledge about these fascinating creatures and shifted the focus from flamethrowers and cave destruction to systematic control and, in the case of *Diaemus* and *Diphylla,* conservation efforts. Additionally, a better understanding of vampire bats has helped to dispel myths and misconceptions about the eleven hundred nonvampire bat species as well as blood feeders in general. We can now spend more time dealing intelligently with our attraction to nature's vampires as well as the unique substance that ultimately led to adaptations like razor-sharp teeth and salivary anticoagulants. That substance is, of course, blood—to many, the source of life. But considering our seemingly innate feelings of attraction and revulsion toward blood, until recently, our relative lack of knowledge about the red stuff made us look positively erudite about vampire bat biology.

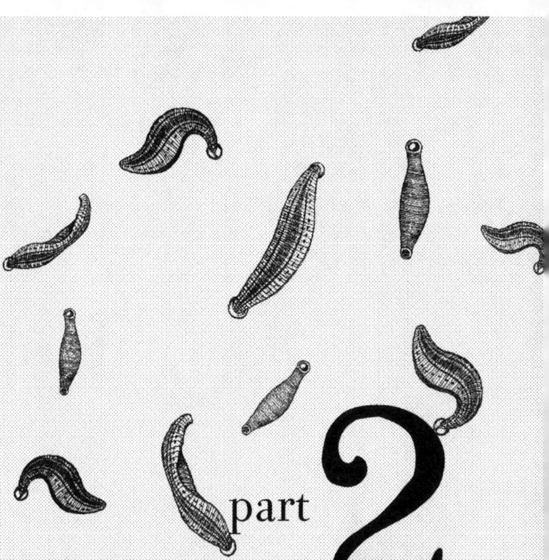

part 2

Let It Bleed

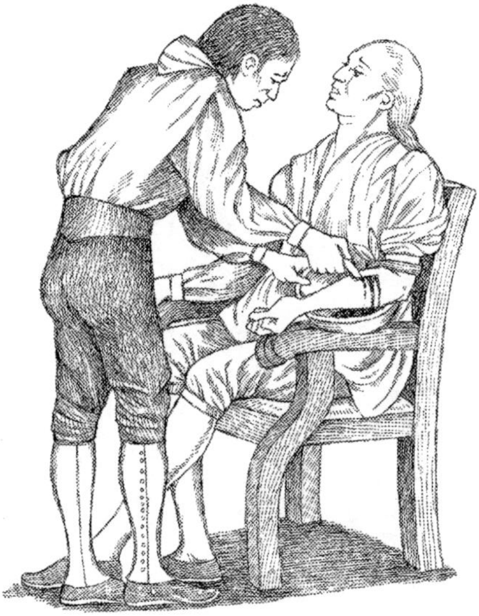

I firmly believe that if the whole *material medica* as now used could be sunk to the bottom of the sea, it would be all the better for mankind—and all the worse for the fishes.

—Oliver Wendell Holmes

Nearly all men die of their remedies and not of their illnesses.

—Molière

Several hours before his death, after repeated efforts to be understood, [he] succeeded in expressing a desire that he might be permitted to die without further interruption.

—Drs. James Craik and Elisha Dick
(December 31, 1799)

4.

EIGHTY OUNCES

On the morning of Friday, December 13, 1799, the first president of the United States woke up with a sore throat. He had been riding at his farm the day before and the weather had been cold and windy—snow giving way to hail and finally rain. To make things worse, although his clothes had gotten soaked, he refused to change out of them before dinner. That evening, George Washington stayed up late, reading newspapers and asking his private secretary, Tobias Lear, to read him an account of the Virginia Assembly's debates on the selection of a senator and governor. Washington, whose voice had become hoarse, took no treatment for what he perceived to be the start of a simple cold.

Later, Martha, who could tell that her husband was starting to come down with something, chided him for coming to bed so late. "It has been my unvaried rule never to put off till the morrow the duties which should be performed today," *reportedly* came Washington's famous reply to his wife.

At around 3 a.m., on December 14, the Founding Father woke with a fever. He also found it hard to speak and was having difficulty breathing. A mixture of molasses, vinegar, and butter was prepared, but Washington began choking violently when he tried to swallow it.

The former president's longtime physician, Dr. James Craik, was called in, but before the doctor arrived Washington sent for his estate overseer, Albin Rawlins, who appeared just after sunrise. Rawlins's medical experience consisted of treating sick livestock, but that didn't stop Washington from ordering the man to bleed him. Although this would seem to be a bizarre request, in Washington's time bloodletting or "breathing a vein," as it was called, was an extremely common treatment, comparable nowadays to popping a couple of aspirin. Still, after preparing the former president's arm, Rawlins hesitated.

"Don't be afraid," Washington told the overseer, who suddenly appeared reluctant about slicing into his master's arm. Soon after, the incision was made, and according to Tobias Lear, "The blood ran pretty freely."

A pint of blood was removed.

In Colonial America (as well as throughout Europe), a bladed instrument called a lancet was commonly used during bloodletting. Lancets were produced in huge numbers and often stored in ornate cases. Their importance can be demonstrated by the fact that England's premier medical journal took its name from the instrument. After tying off the arm at the elbow, the lancet was applied lengthwise to the now bulging vein, which prevented the vessel from being severed in two. A number of spring-loaded me-

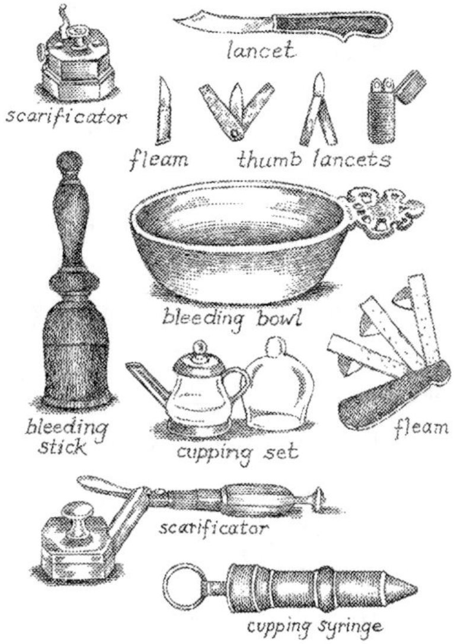

chanical devices (called scarificators) were also concocted to facilitate the drawing of blood, which was often collected in specially designed bleeding bowls. Some of these were quite beautiful, and they were often marked with inner concentric rings arranged in one-ounce increments.

When Washington's wife, Martha, came upon the gory scene she begged that the procedure be stopped, but her husband insisted that the bleeding continue.

"More, more," he said, hoarsely, after complaining that the cut Rawlins had made wasn't large enough.

After the bleeding, Washington's neck was wrapped in flannel soaked in hartshorn ammonia* and his feet were bathed in warm water.

Dr. James Craik had been treating George Washington for over

*This was a type of baking powder, also known as baker's ammonia or ammonium carbonate.

thirty years, and when he arrived, at around 9 a.m., he knew immediately that he was dealing with a potentially fatal illness. He applied a blister of cantharides to Washington's throat,* then instructed his famous patient to gargle with yet another mixture—vinegar and sage tea this time. The results must have been terrifying to those standing by. The former president, who was already struggling to breathe, nearly suffocated on the concoction. Alarmed that Washington could not swallow and was having great difficulty breathing, Dr. Craik sent messengers calling for two additional physicians, Dr. Elisha Dick and Dr. Gustavus Brown. Following this, Craik bled the former president and repeated the procedure at eleven o'clock.

By the time the two additional doctors arrived later that afternoon, Washington had been bled a total of three times. Although Washington's skin was blue when they arrived, to the physicians of his time this would have indicated an improvement in their patient's condition (except for the fact that the man was unable to speak, swallow, or breathe without great effort). Medical men of the day often sought to reduce fever or a high pulse rate by decreasing the volume of body fluids. It appears that as far as a patient's condition was concerned, cool, calm, and cyanotic was preferable to feverish, frenetic, and flushed. Inflammation and fever were not yet recognized as the body's attempt to combat infection, and so patients were typically bled to alleviate the heat, redness, and pain that were thought to result from "abnormal vascular congestion."

According to George Washington Custis (Martha's grandson

*Cantharides is a preparation made from the dried, crushed bodies of blister beetles (family Meloidae). Applied externally, cantharides is a natural irritant and raises serum-filled blisters, once thought to draw sickness out of the patient's body. Also known as Spanish fly, these smashed beetle bits have been touted as an aphrodisiac since the time of ancient Rome. In reality, ingesting even small amounts of this material can cause vomiting, abdominal pain, kidney failure, and death.

from her first marriage), the three physicians held a brief consultation: "The medical gentlemen spared not their skill, and all the resources of their art were exhausted."

What course of treatment had the three doctors agreed upon? Additional bloodletting, and their patient (who had no medical training) agreed with them.

Another full quart of blood was removed, so that by early evening George Washington had been drained of *eighty ounces* within a thirteen-hour period. Much to the dismay of everyone present, Washington's blood flow was so weak (and his blood so "thick") that physicians were unable to induce him to faint, the long-sought end point for most bloodletting treatments.

Deeply distressed at his rapidly deteriorating condition (he was reportedly in great pain and having tremendous difficulty breathing), the doctors administered calomel, a strong laxative and emetic tartar.* These purges were as commonly administered as bloodletting in the eighteenth century, but in this instance, although they had their expected results—"a copious discharge from the bowels"—there was no improvement in the patient's condition.

Alarmed but undaunted, the three physicians kept trying. In an attempt to draw the poisonous humors away from Washington's throat, blisters were raised on his extremities. Additionally, a poultice of wheat bran was applied to his feet and legs, and he was asked to inhale a solution of ammonia and water. Still, the former president's condition continued to worsen. Unbeknownst to his doctors, not only had Washington lost a dangerous amount of blood but the "medications" they prescribed to purge him had, in all likelihood, left their patient *severely* dehydrated as well.

*This mixture of antimony and potassium tartrate was commonly used to induce vomiting.

In desperation, Dr. Dick suggested a tracheotomy (or bronchotomy, as it was then referred to). This was a relatively new procedure that was sometimes employed to treat injuries in which the larynx (voice box) had been crushed.*

Unfortunately, no one will ever know if this procedure could have saved George Washington's life. Apparently, the two elder attending physicians overruled Dr. Dick.

Back at Washington's bedside, it soon became apparent to all (including the great man himself) that the end was near. According to Tobias Lear, by this point, his master "spoke but seldom, and with great difficulty; and at so low & broken a voice as at times hardly to be understood." Washington asked his wife to fetch two wills from his desk, then instructed her to burn one of them (since it had been superseded by the other). He called upon Lear and struggled to make certain that his letters and papers were in order and that his accounts would be settled upon his death.

At around 10 p.m., Lear bent in close as George Washington struggled to voice his final words. "I am going," he said. "Have me decently buried; and do not let my body be put into the Vault in less than three days after I am dead." This seemingly odd statement can perhaps be explained by a widespread fear of premature burial that existed at the time.

After Lear confirmed that he'd understood what Washington had said, the first president of the United States uttered his last words, "'Tis well."

Shortly after that, the most famous man on the planet expired, "without a struggle or a sigh."

*During this procedure, a small midline incision was made in the throat at a level just below the larynx. Next, a half-inch-long horizontal incision was made in the trachea at the level of the third tracheal ring. Following this, a small, hollow cannula was inserted into the severed airway and secured in place by ribbons, which were tied around the neck. This arrangement allowed the patient to breathe, even in the presence of an obstructed larynx. Once the patient had recovered, the cannula was removed and the wounds from the tracheotomy were sutured closed.

George Washington was sixty-seven years old.

Over two hundred years later there is still debate over the specific illness that had stricken the Founding Father. While some have suggested laryngeal diphtheria or a peritonsillar abscess, most experts now believe that Washington suffered from an acute case of bacterial epiglottitis (an inflammation of the leaf-shaped flap that covers the entrance to the trachea during swallowing). Although rare in these days of antibiotics, this malady is potentially fatal since it causes the epiglottis to swell, blocking the airway and rendering the sufferer (as in Washington's case) unable to breathe.

There is *no* debate that the loss of approximately *40 percent* of Washington's blood volume within a thirteen-hour period hastened the great man's demise. For comparative purposes, the American Red Cross generally requires an eight-week period between blood donations of *one-tenth* the volume drained from the former president on what was to be his last day alive.

Eighty ounces.

"What the hell were those guys thinking?" I asked myself, initially.

But as easy as it is to scoff at Washington's physicians and their chosen course of treatment, I know now that I was dead wrong in doing so. These men were, after all, simply trying to save their patient's life, and bloodletting had been the accepted treatment not only for sore throats but also for scores of other maladies since the time of the Mesopotamians, Egyptians, and ancient Greeks (as well as the Mayans and Aztecs).

There was, in fact, a storm of criticism following Washington's death and those comments are quite revealing. One physician claimed that Washington's tonsils should have been scarified, while another suggested that the former president's doctors should have bled him from under the tongue, since that location was anatomically closer to the problem. Other suggestions included rubbing Washington's throat with "warm laudanum" (a mixture of alcohol and opium) and, following this, the application of a bag of warm salt to his neck. Instead of calomel, some insisted that he should have been urged to drink small portions of hot whey, laudanum, or *spiritus volatilis aromaticus,* a mixture of ammonia, carbonate of potash, cinnamon, cloves, and lemon peel.

Rather than condemning Washington's attending physicians, these nineteenth-century armchair quarterbacks have actually exonerated them, for it is unlikely that *anyone* could have saved him using the medical practices of the day. What might be even more difficult to accept is that even if George Washington had lived a century later—in all likelihood his treatment and its outcome (in the absence of antibiotics) would have been *exactly* the same (the only difference being that his physicians probably would have used leeches to draw off his blood).

But why was this so? With tremendous advances in nearly every field of science, why were renowned physicians still bleeding their patients, often to fatal excess, right into the twentieth century? How did this practice of bloodletting come about? Does it ever work and if not, then why are thousands of leeches being used by physicians today around the world?

To answer these questions, we'll need to take a slight detour from blood-feeding creatures to examine our knowledge (or until recently, our lack of knowledge) concerning blood and the circulatory system.

As opposed to our longstanding misconceptions about bats, bad information on blood and the circulatory system didn't start with the Europeans—although they were responsible for carrying this miserable banner for well over a thousand years.

The ancient Egyptians certainly had some vague ideas about

the function of the circulatory system (as revealed in the Smith and Ebers papyri, dated to the seventeenth and sixteenth centuries BCE, respectively). They knew, for example, that there was a relationship between the heart and a person's pulse. Their knowledge was somewhat limited, however, by an inability to differentiate among blood vessels, nerves, tendons, and ureters (the tubes which carry urine from the kidneys to the urinary bladder). This led to confusion about substances like sperm, urine, tears, and blood, although the latter was put to some use. For example, some ancient Egyptians took the blood of a black calf or a black ox, mixed it with oil, then slathered it on top of their heads. Why? To combat graying hair, of course—a sort of Grecian Formula 44 before there were actual Greeks. Maybe that sounds ridiculous today, but the implication is that there was *something* in the blood itself that would restore blackness to the hair.

The word *blood* shows up in the Bible over four hundred times with the earliest mention coming in the Old Testament (Genesis 4:10), right after Cain has murdered his brother, Abel: *"And the Lord said, 'What have you done? The voice of your brother's blood is crying to me from the ground.' "* Later, an even more significant passage concerning blood occurs in Genesis 9:6: *"Whoever sheds the blood of man, by man shall his blood be shed."*

Since the ancient Hebrews believed that the spirit resided in the blood, killing someone was more often referred to as spilling or shedding their blood. It was and is, arguably, the most serious thing we can do to another human. So serious, it seems, that it calls for the spilling of *more* blood from those deemed responsible.

Blood pops up (oozes up?) again in the Old Testament, in Genesis 9:3–4, as God is chatting with Noah: *"Every moving thing that lives shall be food for you; and as I give you the green plants, I give you everything. Only you shall not eat flesh with its life, that is, its blood."* (Note: This passage follows right after God's more popular command that Noah and his sons *"be fruitful and multiply."*)

This notion, that life itself resided in the blood, appears to have led to the requirement that animals slaughtered for food be drained of blood before being eaten.* Apparently, exsanguination spares people from eating the spirit of the animal—which presumably escapes the red puddle sometime *before* the clotting process is completed. Did I mention that this procedure also comes in handy if you're looking for a way to feed your vampire bat colony?

The Hebrews certainly weren't alone in their belief that blood was a special juice, nor were they the only ones who spent considerable time and effort spilling it for those beliefs. So important was this fluid that the offering up of blood was thought by many cultures to be the ultimate way to atone for one's sins, to pay homage to one's god or gods or to cure one's self of various ills. (Countess Báthory serves as a rather extreme example of the latter.) Generally, animals were sacrificed. Calves were popular, possibly because they were easy to lead around and their blood could cover a lot of altar surface, but in far too many cases to fathom human blood was considered to be the ultimate sacrifice—whether it was shed for atonement, revenge, the gods, or a cure.

Scientific knowledge of blood and the circulatory system was, to put it mildly, rather slow to come about. To put it less mildly, most of the early information was dead wrong, yet somehow it lingered in the field of medicine for over two thousand years. Here's how it happened.

Around 400 BCE, Hippocrates proposed (among *many* other things) that the human body contained four substances called humors: black bile, yellow bile (or choler), phlegm, and blood. When the four humors were balanced, the person remained healthy, but any humoral imbalance led to sickness, misery, and

*The ban on consuming blood appears elsewhere in the Bible, such as Leviticus 7:26–27: *"Moreover, you shall eat no blood whatever, whether of fowl or of animal, in any of your dwellings. Whoever eats any blood, that person shall be cut off from his people."*

despair. Accordingly, to the ancient Greeks, it was the *volume* of blood that dictated the health of an individual.* It should come as no surprise, then, that starvation, vomiting, and bleeding were used to treat perceived excesses in humors, while other patients were instructed to gorge themselves when their humoral levels needed boosting.

As adopted by Claudius Galenus (better known in English as Galen) nearly six hundred years after Hippocrates, humoral imbalances were not only used to explain how people got sick but how they got their personalities. Too much phlegm, for example, led to a lack of emotion in the individual—a phlegmatic personality. On the other hand, too much blood led to a "sanguine" or carefree temperament. In this case, nosebleeds, hemorrhoids, and menstruation were looked upon as the body's way of restoring normal blood levels.

Earlier, while working as a physician at a gladiator school in the Turkish city of Pergamon, Galen, the son of a wealthy architect, had glimpses of internal human anatomy, referring to wounds as "windows into the body." The thirty-two-year-old Galen moved to Rome in 160 CE, but the Roman ban on human dissection meant that he would never get to explore what was on the other side of that window. Galen was reduced to making inferences about the human condition by examining animals like macaques (a type of Old World monkey), pigs, and goats. These animals were often dissected alive and in public, and Galen's "hands-on-guts" public demonstrations made him incredibly popular. Unfortunately, although Galen's dissections set him apart from more traditional physicians (who employed a distinctly less hands-on approach), his reliance on inference, conjecture, and his own

*We know now, of course, that in many instances it is the presence or absence of pathogens (i.e., disease-causing organisms) in the blood and elsewhere that determine one's health.

imagination often led to conclusions about the human body that were dead wrong.

Galen eventually became the personal physician of Emperor Marcus Aurelius and later his son Commodus.* Although Galen was much more interested in the central nervous system than he was in the circulatory system, he did prove, among other things, that blood, not *pneuma* (an airlike spiritual essence dreamed up by the ancient Greeks), traveled through arteries.

On the other hand, Galen had no real concept of blood circulation. He believed that blood ebbed and flowed like the tides, with venous blood originating from and returning to the liver.† Unwilling to abandon the concept of *pneuma,* Galen proclaimed that blood within the heart passed through invisible pores in the wall separating the heart's chamberlike ventricles. After mixing with the *pneuma,* the blood was then distributed to the body.

Granted, as far as circulatory system basics go, figuring out that blood and not air was carried in arteries was significant, but Galen's deeply flawed concepts of human anatomy and physiology would have a serious and long-lasting effect on the field of medicine—*especially* with regard to the circulatory system. As previously mentioned, Galen's overarching ideas on the human body were generally extensions of those proposed by the ancient Greeks, and these mistake-laden views came to completely dominate the field of medicine. Not only did Galen's take on medicine and anatomy remain pervasive for fifteen hundred years, it remained *unchallenged.* According to Bill Hayes, the author of *Five Quarts—A Personal and Natural History of Blood,* "In the early Middle Ages, church leaders declared his work to have been divinely inspired and thus infallible." Rather than experimenting or dissecting specimens (and thereby bringing down upon themselves

*Commodus was portrayed by actor Joaquin Phoenix in Ridley Scott's film *Gladiator.*
†One function of the liver was thought to be the conversion of tiny particles of food into blood.

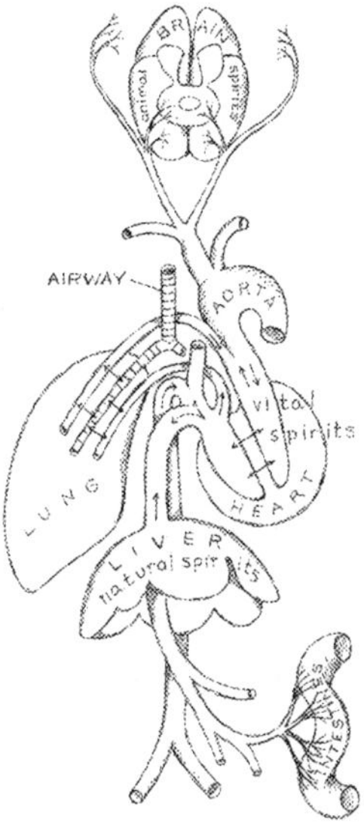

the serious and often deadly wrath of the church), the disciples of "Galen the Devine" simply deferred to their long-deceased master and his stance on any given medical topic. Anything else would have been blasphemous.

How did bleeding become such a popular therapeutic tool? What was it that compelled the most learned physicians of their day to drain their patients until they were cold, blue, and unconscious?

During ancient times, bleeding was generally thought to rid the body of evil spirits. Later, once the concept of balancing humors became accepted, regularly scheduled bouts of bloodletting were prescribed in much the same way that a balanced diet and exercise are extolled today. For example, fever and headache were thought to be symptoms of excess amounts of blood ("plethoras") and called for immediate drainage. Galen considered blood to be the most important of the four humors. (Is anyone shocked that it edged out both shades of bile and survived a late run by the cigarette smoker's favorite–phlegm?) He used his knowledge and expertise to write a series of books that soundly trounced his critics–especially those who denounced his bloodletting techniques.

After Galen, the significance of blood as a humor gained even greater import, especially when folks determined that there was a *slight* problem with black bile–it didn't exist.* Blood, on the other hand, was real and it could be tapped by any number of methods. Galen and his contemporaries used a metal scalpel called a phlebotom (from the Greek words for "vein" and "cut") to make a small venous incision through which a pint or so of blood would be drained. Influential physicians drew up complex charts based on parameters like the seasons, tides, and weather to calculate the amount of bleeding to be done.† Similarly, Hebrew and Christian writings also prescribed the best days for bloodletting to take place.

Most historical accounts of mammalian circulation emphasize the work of William Harvey, who, in the early seventeenth century, used scientific methodology to prove that blood did not ebb and flow like the tides. Instead, the heart pumped it around the body in a pair of loops, one to the lungs and back (the pulmonary circuit), and one to supply the body and its tissues (the systemic circuit).‡ Before Harvey's discovery, the general belief was that

*Black bile was supposedly produced by the spleen and was thought to be responsible, among other things, for the dark coloration of bodily substances, like blood and feces. Differing levels of black bile were also used to explain why some people had darker skin than others.

†In 1462, a bloodletting calendar was the second medical text to be mass-produced using Johann Gutenberg's revolutionary printing press. This was some eight years after the first Gutenberg Bible.

‡Less well known is the fact that the Arab physician Ibn al-Nafis (1213–1288) had described much of this dual-circuit pump four hundred years earlier.

disease-laden "bad blood" had a tendency to pool in the extremities where it would stagnate. Bloodletting, therefore, was a way to eliminate the bad blood. Unfortunately for George Washington (and countless other patients), physicians mistakenly thought that the often copious amounts of blood they drained would be replaced within a very few hours by new, healthy stuff. As Dr. Craik and his colleagues would never learn, this just wasn't the case.

Beginning in the seventeenth century, bloodletting was not generally undertaken by physicians or surgeons (the top two rungs of the medical practitioner's ladder, respectively). Procedures such as therapeutic phlebotomy, leeching, and even minor surgeries were usually carried out by a lower class of medical personnel, the barber-surgeons (who were themselves a rung or two higher than midwives). Barber-surgeons were the descendants of the bath men who toiled in medieval bathhouses. Both had duties that included shaving, cutting hair, bleeding patients, administering enemas, and changing wound dressings. During wars, some barber-surgeons traveled with their respective armies—treating fractures and probing for bullets. They became the first military surgeons. Back home, they advertised their talents with a striped barber pole outside their establishment—the red stripes signifying blood, blue stripes were veins, and white stripes represented the gauze bandages they used to stem the bleeding. The pole itself was a symbol of the stick that patients would grip tightly as they were being bled and the ball atop the pole signified the blood collection basin (and the container they used to hold leeches).

Barber-surgeons "breathed veins" to treat all serious maladies and any number of lesser complaints, from asthma and bone fractures to drunkenness and pneumonia. Women were bled to *reduce* menstrual flow and "lunatics" were drained to treat mental illness. Even drowning victims were bled!

In modern times, with the remarkable medical advances that

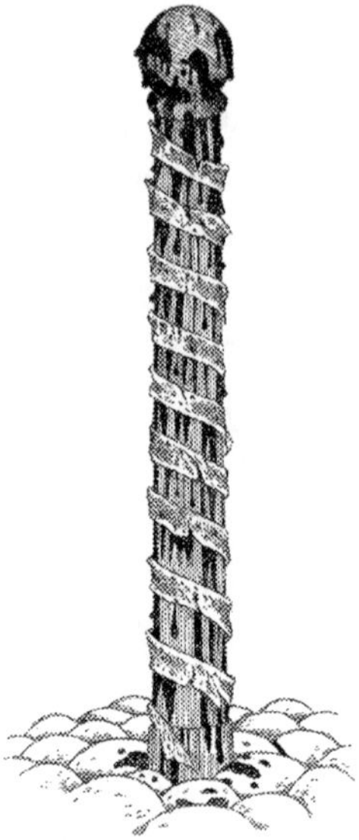

we see on an almost daily basis, it's easy to overlook the fact that in many ways medical research was relatively stagnant from the time of the ancient Greeks until the first decades of the twentieth century.

One of those responsible for attempting to revive experimental medicine was Andreas Vesalius. Born into a family of Belgian physicians, Vesalius received his doctorate in 1537 from the University of Padua, where he soon became the chair of surgery and anatomy. There, as in the other early medical schools, Galen's massive literary output served as the basis for all relevant courses and their syllabi. But rather than blindly accepting Galen's well-worn teachings, Vesalius took a new and dangerous approach. He employed dissection in his classroom and preached a hands-on approach to his students. Fortuitously, a sympathetic judge gave Vesalius access to the corpses of executed criminals. The young anatomist not only studied their anatomy but also produced a set of remarkable and highly detailed anatomical diagrams, which were included in his seven-volume *On the Fabric of the Human Body.* It was his masterwork and it hammered Galen's inaccurate and erroneous views on anatomy into the ground like so many tent pegs. Using cadavers, Vesalius disproved Galen's concept of invisible pores in the heart. He also demonstrated that the human heart had four chambers (not three) and that half of the body's major blood vessels did *not* originate in the liver (as described by Galen). Additionally, Vesalius clearly showed that the liver itself was not the five-lobed organ that Galen claimed it was.

Understandably, Vesalius (who was not yet thirty) upset many of the Galen faithful by dismantling so many of their master's long-held claims. One outraged Galenite went so far as to publish a paper in which he asserted that the work of Vesalius didn't prove Galen wrong, it simply indicated that the human body had *changed* since Galen's time.

Vesalius died in 1564, after his ship was wrecked returning from a pilgrimage to the Holy Land. A long-held rumor that he had fled to the Mideast to escape the Inquisition (after dissecting a "corpse" whose heart suddenly started beating) has been discredited.

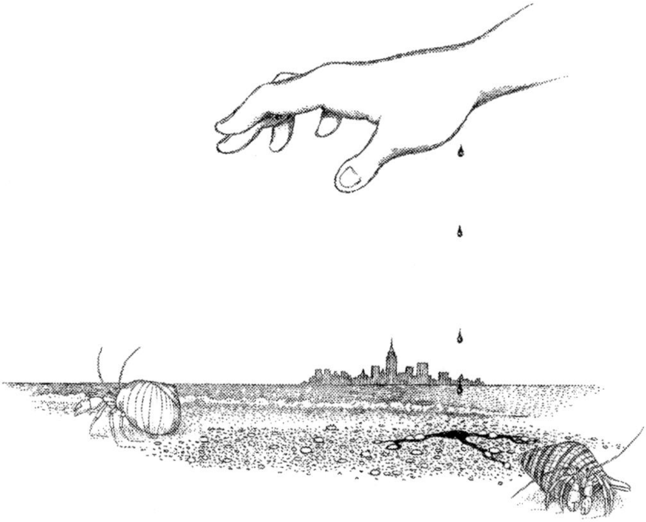

Blood is a very special juice.

—Mephistopheles, speaking to Faust

5.

THE RED STUFF

One day while I was rummaging around in a tidal flat on the South Shore of Long Island, a stray bit of metal ribbon sliced a neat new crease in my wrist. I was around ten years old at the time and I remember staring in silent fascination as the blood welled up and then began to run out of the half-inch cut and down my arm.

My mother, who had been standing nearby having a smoke with my aunt Rose, must have come up behind me to see what I was doing. (I guess I hadn't moved from where I was squatting for nearly a minute, and in my mother's mind this was clearly reason enough for alarm.)

"Are you out of your *friggin'* mind?" she screamed, nearly sending me sprawling into the mud.

"Oops," I said.

Realizing immediately that I'd have to defuse the situation, I pointed to my wounded wrist. "It doesn't hurt, Mom," I remarked cheerfully.

"You *are* out of your friggin' mind," my mother shrieked, grabbing me by the nearest nonbleeding appendage and hauling me toward Aunt Rose.

This is not going to be pretty, I thought.

I should explain that when I was a kid, I had something like eight Aunt Roses. As you can imagine, telling them apart was something of a dilemma ("That's Aunt Rose *DiMango,* not Aunt Rose *DiDonato*!"). So, inspired by the Peterson Field Guide series, I approached the problem by developing a set of identifiable field characteristics based on traits like "total body length" and "facial mole placement" (an original creation of mine). Within the Aunt Rose group, individuals ranged in height from a kid-friendly four feet eight inches on up to a towering sixty inches. This one was a midsized Aunt Rose, easily identifiable however, by her miniature poodle, Fifi, as well as her unique talent for combining body language and swearing into a kind of interpretive Italian dance. And as my mother dragged me toward the sidewalk, bleeding and ear tweaked, I couldn't help noticing that my aunt had already initiated some preemptive hand gestures (holding an index finger to her temple, thumb extended skyward).

I don't remember too much about the cleanup—although I do recall getting blood on one of Aunt Rose's bathroom statues. ("Gracie, the sa-na-va-*bitch* is bleeding all over the Virgin Mother!")

Many people have been and continue to be intrigued by blood, while others (like my mom and at least five of my aunt Roses) are—how should I put this?—somewhat less than intrigued by it. But whether you were fascinated by blood or repelled by it, as a child in the 1960s we were all starting to see quite a bit more of the red stuff. The assassinations of President Kennedy and his brother Senator Robert Kennedy, the student deaths at Kent State, and the Manson murders were beamed right into our living rooms, their visual horror intact. Color news footage from the Vietnam War seemed to take a hard turn toward graphic around the time that Sam Peckinpah's *Wild Bunch* and Arthur Penn's glamorized versions of *Bonnie and Clyde* were going down in slow motion—in a blaze of explosive Technicolor squibs.

Forty years ago, the sight of all that red was shocking. Today, some of us are still repelled by blood. Others are titillated by it (think of the mountains of money brutal gore-fests like the film *Saw* have taken in). Still others have become acclimated to it. Similar to the way our bodies learn not to respond to inconsequential stimuli (you don't feel your socks once you pull them on, do you?), we have adapted to the sight of blood with a corresponding decrease in sensitivity.

So what is blood, exactly? One answer is that it's food for the creatures inhabiting this book. Because of that, I don't feel badly about taking a few detours to explore the substance a bit.

An anatomist might start by describing blood as a connective tissue—just like bone, cartilage, tendons, and ligaments. Confused? Well, not for long, once you figure out what it takes to be considered a card-carrying connective tissue. Tissues are accumulations of different types of cells (along with their matrix, which is the noncellular medium that surrounds them). At an organizational level, they're one rung on a kind of hierarchical ladder that characterizes all living things. In this regard, tissues are a rung above cells and a rung below organs, which are structures that are composed of several *different* tissues. Taking this hierarchy a bit further, several organs, working together, form an organ system, and organ systems combine to form an organism. Going in the opposite direction on our ladder, cells are made up of subunits called organelles (such as the oft-memorized mitochondria), and organelles are composed of biochemical units like proteins and lipids. And so on.

Okay, back to tissues.

The next requirement for a tissue is that its cells work together in some particular function. For example, nervous tissue is composed of neurons and a support team of glial cells—each contributing in specific ways to the functioning of the nervous system.*

Connective tissue is characterized by being composed of relatively few cells surrounded by a significant amount of noncellular matrix. As a result, connective tissue cells are generally not in contact with each other. Think of them as bricks in a wall, with the matrix being the mortar that surrounds them and binds them to each other. The matrix is also what gives connective tissues their physical properties. For example, the hardness of bone comes from calcified bone matrix, not from the bone cells (osteocytes) themselves.†

*Besides connective and nervous tissue, there are two additional tissue types: epithelial tissue, which covers surfaces and lines hollow structures, and muscle tissue, unique for its ability to store chemical energy, then convert it into the energy of motion as the tissue contracts in size.

†The flexibility of cartilage, another type of connective tissue, comes from a gel-like matrix, which is basically bone matrix containing bendable protein fibers instead of calcium and mineral salts. Tendons and ligaments, which connect muscle to bone and bone to bone, respectively, are also structures composed of connective tissue, and they get their strength from the tough, wirelike fibers found embedded within their matrices.

In blood, the matrix is called blood plasma and it's neither solid nor gel—it's a liquid (composed primarily of water)—and this property is just as important functionally as hardness, flexibility, and strength are for the other types of connective tissue. The reason is that plasma acts as the transport medium for blood cells, as well as tiny cell fragments called platelets that are involved in blood clotting. Additionally, many other important substances are carried within the plasma in a dissolved state, including nutrients, vitamins, hormones, waste products, gases, and ions.

Pumped out of the heart by the muscle-bound left ventricle, the force exerted by the blood on the inside of the vessels it passes through is known as blood pressure. When the left ventricle contracts, expelling its arterial blood, the blood pressure increases. This produces the higher number (known as the systolic pressure) in a typical blood pressure measurement. As the left ventricle empties, relaxes, and begins to fill again, the blood pressure drops, producing the lower diastolic pressure.*

As the arterial blood nears its destination, it moves from arteries to smaller arterioles and finally to miles and miles of microscopically tiny (and thin) capillaries. These mini-vessels form dense, netlike beds around organs and other structures. During this passage to smaller and smaller blood vessels, the blood pressure drops significantly. To understand how this pressure drop takes place, visualize the water traveling through a garden hose. Now imagine that the far end of that hose begins to split into smaller and smaller tubes, each of those tubes splitting again and again—until the end of the hose has been divided into a million tiny branches. The pressure of the water in any one of these

*Like other pressure measurements (e.g., barometric pressure), blood pressure is measured in millimeters of mercury, so that a systolic pressure of 110 represents a force applied to the blood vessels' inner walls equivalent to one that could raise a thin column of mercury 110 millimeters in height (were that column of mercury sitting at sea level in a U-shaped glass tube).

branches would be far less than the original pressure. Basically, this is because of the tremendous increase in the total combined area inside the millions of branching tubes (as compared to the area inside the original hose). This is also the reason why the water pressure drops when everybody decides to take a shower at the same time.

But not only are these low-pressure capillaries incredibly small, their walls are so thin that once the blood gets to its destination, nutrients and oxygen contained within the blood plasma simply diffuse through the capillary walls to supply the surrounding tissues and their cells.*

Metabolic waste products and carbon dioxide move in the exact opposite direction (from the tissues *into* the blood) via the same process. Once within the capillaries, they start their journey back to the heart, passing through increasingly larger vessels (venules leading into veins) before entering the heart's right atrium. Unfortunately, this low-pressure blood sometimes has a hard time returning from places like the legs and feet—since it must overcome the considerable force of gravity. Under normal conditions, venous return is aided by a series of one-way valves and something called the skeletal muscular pump. In places like the calf muscles, involuntary muscular contractions squeeze the veins traveling through them. The low-pressure blood within the compressed veins shoots upward against gravity and back toward the heart, while the valves prevent backflow of the blood toward the feet. You can actually see this process in action by sitting cross-legged and watching your calf muscles. The irregular twitching you see is the skeletal muscular pump in operation.

*Oxygen (which is actually carried around inside red blood cells) and nutrients diffusing in this manner are said to be following a concentration gradient, moving from areas where they're in a higher concentration (the blood) to areas where they're less concentrated (e.g., oxygen- and nutrient-starved tissues and cells).

Unlike large complex creatures (like us), microscopic organisms don't have blood or a wildly intricate circulatory system (and they don't have organs or organ systems either). For the majority of species on this planet, things are generally much simpler.

Picture how easy it would be for a single-celled amoeba to exchange gases with its environment. With only one cell to supply, oxygen and other incoming substances are obtained *directly* from the environment. Although some of this transport requires energy expenditure, in many instances, incoming and outgoing material and gases simply follow concentration gradients across the organism's thin cell membrane. Now imagine a creature shaped like a ball—composed of millions of cells. How would the cells toward the center of the ball get their nutrients and oxygen or rid themselves of waste? Although you might be able to think up several ways that this *might* occur—the solution that evolved for many creatures here on earth consists of a muscular pump (the heart), a remarkably intricate vascular transportation system, and a unique and versatile tissue carried within it: blood.

Before we get too carried away with just how cool our circulatory systems are, you should know that there are some relatively large organisms that get by just fine without elaborate circulatory systems. Insects, for example have low-pressure systems that are termed *open circulatory systems* because they don't form a completely closed loop between the body and heart. Hemolymph (the arthropod equivalent of blood) is circulated through the body by a series of heartlike dorsal pumps and by movement of the insect's body. The hemolymph moves through vessels that eventually lead to open sinuses called hemocoels. Here, the surrounding internal organs are literally bathed in the nutrient-bearing fluid, which

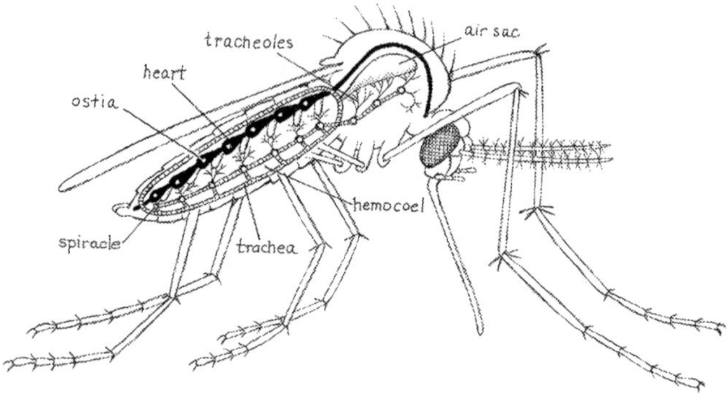

eventually percolates back and reenters the hearts through tiny valves called ostia.

The key here is that unlike animals like vertebrates, insect circulatory systems aren't involved in transporting gases like O_2 and CO_2 or exchanging those gases with body tissues.* A mosquito's oxygen requirements are met through a series of openings (spiracles) along both sides of its thorax and abdomen. Air passes from the environment into the spiracles (which can also be closed to prevent water loss), and then through a complex of tubes called tracheae. The tracheae get smaller and smaller, finally branching into microscopic vessels called tracheoles, through which the air finally arrives to supply the tissues and cells.

This system works just fine for small creatures, but there are

*Like vertebrate blood, hemolymph does carry nutrients, hormones, and metabolic waste. It also functions in clotting and the immune response.

limitations. For example, tracheal respiration is probably a key reason why mosquitoes (and other insects like bed bugs) aren't a whole lot bigger in size than they are. Larger animals are composed of too many cells to be efficiently supplied by this type of respiratory system.

Some of you might be saying, "Wait a minute, what about those pictures of ancient dragonflies with three-foot wing spans? How did *they* get enough oxygen?"

The answer is that there is evidence from the Carboniferous period (290–360 million years ago) that widespread forests and lush plant life resulted in a higher percentage of atmospheric O_2 than exists in today's atmosphere. This extra O_2 was apparently enough to support larger species of insects that employed the same tracheal system as their smaller modern-day cousins. Still, during the Carboniferous or even during the age of dinosaurs (when some creatures reached gigantic proportions), there is absolutely no evidence of Mothra-sized insects (or, for that matter, the twin fairies who controlled them).

To help carry out myriad circulatory functions, the average human has between 1.2 and 1.5 gallons (4.5 to 5.7 liters) of blood in their body at any given time. Blood plasma constitutes 55 percent of that volume, while cells (red and white blood cells) and platelets (classified together with cells as "formed elements") make up the remaining 45 percent. Water makes up about 92 percent of blood plasma, with dissolved stuff making up the remaining 8 percent. The majority of these solutes are proteins produced in the liver.

To be considered a tissue, blood needs to be composed of several cell types—and it is. Blood cells (or corpuscles) come in two flavors: erythrocytes (red blood cells) and leukocytes (white blood cells). Erythrocytes (from *erythrós,* the Greek for "red") are by far the most numerous, making up over 99 percent of blood

cells. Function-wise, they're very much like the old Kentucky Fried Chicken (as opposed to the new KFC). That is "they do one thing and they do it right." The one thing they do is carry oxygen, and they do it right because each erythrocyte is literally stuffed full of an iron-containing pigment molecule called hemoglobin. Hemoglobin acts like an oxygen magnet, picking it up where it's plentiful (like in the lungs right after you take a breath) and dumping it off in places where it's in short supply (like tissues, whose cells require a constant supply of O_2 and nutrients). Hemoglobin is so effective at carrying O_2 (compared to say, water) that without it a person would need to have *seventy-five gallons* of fluid circulating in their body to carry the required O_2. And while this would certainly be an exciting development for the fabric industry (bathing-suit sizes would reflect the number of yards of material used to make them), it probably wouldn't be much fun for anyone else.

There is, however, a downside to hemoglobin's efficiency and this is related to the fact that it's even more strongly attracted to carbon monoxide than it is to oxygen. This means that hemoglobin binds to the potentially deadly gas even when there's oxygen present—a property that makes carbon monoxide *extremely* toxic even in small amounts. With the body's hemoglobin tied up with carbon monoxide, the tissues in the brain quickly become starved for oxygen. This leads to a loss of consciousness followed soon after by severe brain damage and death.*

All right, back to blood cells.

Erythrocytes are tiny—only about one three-thousandths of an inch in diameter—and each is shaped like an old rubber stickball being squeezed from opposite sides (nonstickball players often refer to this shape as a biconcave disk). Red blood cells are so focused on their single-minded, oxygen-carrying quest that when mature they lack nuclei or any of the organelles many of us once committed to the rote memorization region of our brains. Erythrocytes are formed in red bone marrow and enter into circulation at a rate of two million per second (which means that they're destroyed and recycled at the same rate by the spleen and liver). There are so many erythrocytes that if they were laid side by side in a single layer they would cover about thirty-five hundred square yards. This provides an incredible amount of surface area for oxygen to cross into tissues.

White blood cells, on the other hand, are a different and altogether more diverse lot than erythrocytes. For starters, they *have* a nucleus and they do not contain hemoglobin (so they don't carry oxygen). Leukocytes have been divided in two major groups

*Hemoglobin's high affinity for carbon monoxide, coupled with the fact that the gas is odorless, are two major reasons why several carbon monoxide detectors are *an absolute must* for every home. In fact, if you're reading this and *don't* have a CO detector, put this book down and go buy one (or order it online). It's *that* important.

(granulocytes and agranulocytes) based on whether or not their cytoplasm (sort of like a matrix inside each cell) looks grainy when stained for viewing under a microscope.*

Functionally, some white blood cells (neutrophils and macrophages) are like blood cell versions of amoebas. Highly energetic, macrophages can usually be found scarfing up microscopic material through a process known as phagocytosis. Rather than looking for food, wandering phagocytes (aka free macrophages) circulate through the body, seeking out foreign microorganisms like bacteria and fungi, and gathering in great numbers at sites of infection. Others (fixed macrophages) remain stationary, in places like lymph nodes or tonsils. Like soldiers assigned to guard a fort, fixed macrophages stand by, ready for action if trouble shows up. This concept of forts or outposts is precisely the reason tonsils aren't removed nearly as frequently as they were when I was a kid.

Upon encountering an invader (recognized by foreign proteins embedded in its cell wall, or by the specific chemicals it gives off), the phagocyte wraps its pseudopods around the microbe, drawing it inward. Inside the phagocyte, ingestion soon gives way to digestion as the foreign microbe is imprisoned within a membranous sac containing a nasty stew of lethal enzymes, bactericides, and strong oxidants. In most cases, the result of this chemical onslaught is a breakdown of the invader's cell wall, followed by a toxic bath and, eventually, death. Any debris that remains is ejected from the phagocyte through a process called exocytosis.

*Blood can be separated into its three major components by spinning it in a centrifuge. A band of erythrocytes (making up 44 percent of the blood volume) will settle to the bottom of a centrifuge tube, with a thin band (1 percent) of leukocytes and blood platelets (known as the buffy coat) sitting above that. The yellow-colored plasma rests on top, making up the greatest portion (55 percent) of the blood volume. By measuring the hematocrit, the proportion of the spun blood composed of red blood cells, doctors can test for conditions like anemia (a decrease in red blood cells) or polycythemia (an increase in red blood cells). For those of you who may be wondering, the blood is prevented from clotting during this procedure by spinning it in heparinized tubes.

Unfortunately, things aren't always so easy for the phagocyte, and as we all know, the good guys don't always win. Pathogenic (disease-causing) microbes have evolved some tricks of their own and many of these organisms have been at it far longer than our immune system has been around to evolve countermeasures. Pathogens like *Staphylococci* bacteria produce their own toxins, which can kill phagocytes. Other invaders, like the AIDS-causing human immunodeficiency virus (HIV), evolve so rapidly that their constantly changing surface proteins are difficult for our immune systems to recognize. Alternately, some invaders, like the tubercle bacillus, are resistant to the phagocyte's usually deadly chemical bath. These pathogens, responsible for the respiratory disease tuberculosis, are taken into the phagocyte. There they multiply within the bag of toxins they've been enclosed in, only to burst forth like Ridley Scott's *Alien,* to kill the phagocyte (and traumatize any phagocytes that might be standing around nearby). Similarly, HIV can hide in these long-lived white blood cells, sometimes emerging after years of dormancy, as if from microscopic Trojan horses.

Some types of white blood cells go about defending the body in a very different way. Leukocytes (like basophils and other connective tissue cells called mast cells) function in the body's inflammatory response. Inflammation is actually a bodily reaction to foreign invaders or tissue damage. During this process, the infecting agent or damaged tissue is partitioned, diluted, and destroyed.

So how does this work?

After getting the call, inflammation-initiating cells release chemicals (like histamine and prostaglandins) that cause blood vessels in the affected area to dilate (increase in diameter) as well as become more permeable. Dilation allows increased blood flow to the site, blood containing oxygen, nutrients, temperature-rising compounds called pyrogens, and an abundance of phagocytic cells. This influx causes the affected area to appear red and even hot to the touch. The increased blood vessel permeability allows plasma (and the phagocyte cavalry) to leak out of the vessels and into the surrounding damaged tissue. The regional swelling that characterizes inflammation is the result.

Macrophages at the site of an inflammatory response to pathogens hunt down and engulf infectious invaders that they encounter. As the battle rages on, millions of them make the ultimate sacrifice and they are posthumously awarded the title of Laudable Pus. Other macrophages get enlisted in the less-popular Cleanup Crew (in which case they get to clean up the laudable pus).* Adding to all this excitement, sensory nerve endings, stimulated by the weird chemicals and the swelling, produce the sensation of pain.

Unfortunately, leukocytes and other protective cells are also responsible for some types of allergic reactions. Most commonly, this "hypersensitivity" results from the body's mistaken (and sometimes life-threatening) attempts to protect itself from nonharmful substances like pollen or dust. Responding to these allergens as if they were a real emergency, basophils and mast cells release their inflammation-promoting chemicals—this time though, in places like the eyes and airways of the lungs (where the allergens have landed).

In far more serious situations, the body's immune system attacks its own joints (rheumatoid arthritis), transplanted organs, or tissue grafts. To prevent extensive tissue damage or transplant rejection, patients sometimes take immune suppressor drugs. One of the most successful has been ciclosporin (or cyclosporine)—a substance originally isolated from a Norwegian soil fungus. It works by decreasing the activity of T cells (discussed below). Although immune suppressor drugs are often taken for extended

*For documentary footage of a large, alien macrophage, I recommend watching the 1958 film *The Blob* (starring Steve McQueen). By doing so, you can get a pretty good idea of how phagocytosis works.

periods of time, the dangers of attenuating one's own immune system should be readily apparent.

Some leukocytes (killer T cells) recognize foreign surface proteins (antigens) and attack any microscopic organisms that "present" them. Other leukocytes (plasma cells, which develop from leukocytes called B cells) produce zillions of tiny bits of protein known as antibodies. Antibodies fit like highly specific keys into locklike receptors on the antigen's surface. The unfortunate bearer of this gaudy antigen/antibody complex either wobbles off to die or is marked for death in much the same way that a guy with toilet paper stuck to his shoe is marked for ridicule.

Other leukocytes (the oddly named helper T cells) function by helping this immune response to occur, while suppressor T cells throttle down the immune response once the battle is over.

Oh yes, before I forget, some plasma cell precursors called memory T cells stick around the circulatory system once things quiet down. Generally, they're dormant, but memory T cells are quite capable of jacking up the immune response at a moment's notice—should the same foolhardy antigen bearer show up again.*

And one more thing: inoculations for childhood diseases (like mumps) and regional diseases (like yellow fever) work on this principle as well. In many cases, dead or harmless versions of specific pathogens are injected into the body, which manufactures antibodies to combat the perceived threat. Additionally, memory cells stick around to quickly crank up the immune response should the real pathogen ever show up.

*This is pretty much why you don't catch the same strain of flu twice and also why you need a new flu shot every year: new viral strains evolve new surface proteins (antigens) that go unrecognized by memory cells or circulating antibodies.

Unfortunately, much of what we know about blood was determined only within the past hundred years or so. Sometimes, though, it was a lack of technology, and not blind devotion to old views, that constrained early researchers. Although the concept of humors would not be extinguished completely for nearly four hundred years, soon after the 1628 publication of William Harvey's landmark book on circulation, *Anatomical Studies on the Motion of the Heart and Blood,* some physicians began to wonder if the benefits of getting someone else's blood into a patient's circulatory system might be greater than draining blood from that person—especially in cases where there had already been significant blood loss.

Dr. Richard Lower performed what is generally considered the first successful blood transfusion in 1666. Using tubing constructed out of goose quills, he connected the carotid artery of one dog (the donor) to the jugular vein of another (the recipient) that he had bled previously to near death. Reportedly, the recipient was resuscitated in a nearly miraculous fashion.

A year later, encouraged by Lower's results, a Parisian, Jean-Baptiste Denis, used a similar device to infuse about eight ounces of calf blood into the arm of a mentally disturbed man by the name of Antoine Mauroy. Denis, like other researchers of his day, believed that blood carried in it the essence of its owner's personality. The rationale for the infusion, therefore, was that the "mildness" of the calf's blood might mellow out Mssr. Mauroy, who police thought was spending a bit too much time running around naked, setting house fires, and beating his wife.*

Tied to a chair, Mauroy was first bled (presumably to eliminate bad blood while freeing up a little space). Then he received

*Back then, people ascribed to the concept of vitalism, in which all creatures possessed an inner force responsible for their specific traits (like bravery in lions or a lust for gold in those working in the insurance industry).

about six ounces of calf's blood, which was introduced through a metal tube. Mauroy complained about some initial burning in his arm but otherwise showed no serious effects. After a short nap, the patient began singing and whistling—which many of the onlookers who had gathered to watch the procedure actually preferred to getting beaten up by Mauroy or having him torch their houses.

Two days later, encouraged by the results, Dr. Denis injected even more calf blood into Mauroy, but this time the results were quite a bit more dramatic. Reportedly, the patient began sweating, and soon after, he complained of severe pain in his lower back (near his kidneys, claimed Denis). Nearly choking to death, Mauroy vomited his lunch and soon after began urinating gobs of black fluid.

Today, physicians would have immediately recognized that Mauroy had suffered a serious reaction because of the extreme incompatibility of the nonhuman blood he had received. The unfortunate man's immune system had in fact mounted a multipronged attack against the foreign blood and the results had nearly killed him. This being 1667, however, the interpretation was somewhat different. To Dr. Denis, surely the vomiting and coal-colored liquid Mauroy continued to urinate for days were proof that the man's madness had been eliminated. After all, the feverish and bedridden patient wasn't nearly as manic as he'd been previously—in fact, he wasn't speaking or moving much at all.*

Several months later, any enthusiasm over the potential benefits of blood transfusion was shattered when one of Denis' patients died. The English, who believed that Denis had not only stolen their transfusion techniques but also their limelight, went out of their way to discredit the Frenchman—as did some of his own rival countrymen. Denis tried to defend himself, but the roof fell in when his highest-profile patient, Antoine Mauroy, also died. After a brief respite, the man had reportedly resumed his wild and brutal ways, and this (as it was later discovered) prompted his wife to employ a bit of creative chemistry. Madame Mauroy began adding arsenic to her husband's diet, but for some reason she failed to mention this fact when she and her husband approached Dr. Denis, asking him to perform a third transfusion. Shaken by his former patient's appearance, Denis declined, but when Mauroy dropped dead several days later, the physician found himself charged with murder. Denis was eventually exonerated, but the uproar surrounding the case (as well as transfusion-related deaths elsewhere) sounded the death knell for human

*Mauroy's dark-colored urine probably resulted from the chemically altered hemoglobin, released by transfused red blood cells blasted apart by the man's immune system. Perhaps owing more to luck than anything else, Mauroy began to recover and Denis reported that on the very first day his patient was able to go to confession, he began urinating normally.

blood transfusions and any related experiments. Two years later, the procedure was banned in France and soon after in England. Additionally, a pair of transfusion-related fatalities in Italy led to a denouncement of the procedure by the pope. Public outcry, capped with a papal denouncement, resulted in a silence that would last for the next 150 years.*

In 1818 gynecologist James Blundell, in an attempt to reduce the large number of deaths associated with postpartum hemorrhaging, performed what is considered to be the first human-to-human transfusion. He withdrew blood from a donor and injected it into a blood vessel in the arm of the donor's wife. From his earlier work on animals, Blundell recognized the importance of eliminating air from the syringe before injecting the blood and also the necessity of performing the transfusion quickly, before the blood had a chance to clot. Survival was a hit or miss affair, and Blundell's first four patients died, not only because they were in a weakened state already but because the well-meaning physician had no knowledge of blood typing or modern anticlotting agents (like heparin). The use of crude, nonsterilized tools only added to the problem.

In 1901, Dr. Carl Landsteiner, an Austrian pathologist, revolutionized the ground rules for blood transfusion after discovering the ABO blood groups.† Simply put, red blood cells (like the foreign microorganisms we saw earlier) have specific surface proteins (antigens) embedded in their cell walls. During transfusions, if the surface proteins on the donor's red blood cells are different from those of the recipient (e.g., antigen A as opposed to

*Strangely, blood was not the liquid of choice during other early attempts at transfusion. According to the American Red Cross, ale, wine, and milk were used. As late as the mid-nineteenth century, physicians injected patients with milk to treat cholera, believing that the "white corpuscles of milk" would convert into the red corpuscles of blood. This wasn't as strange an idea as it might sound, since there are many similarities between the two liquids.

†Landsteiner won the Nobel Prize for his work in 1930.

antigen B), then the erythrocytes in the donated blood will be attacked by leukocytes (or antibodies) circulating in the recipient's blood. This attack by the recipient's immune system kills the donated erythrocytes through a process called hemolysis—literally, "blood cutting" (a process that resulted in the coal black urine seen in the unfortunate Antoine Mauroy). It also leads to a dangerous form of erythrocyte clumping called agglutination, which can clog small blood vessels, sometimes leading to serious medical problems like strokes.

Basically, then, any human-to-human transfusions performed before the discovery of ABO typing were really just crapshoots. And animal-to-human transfusions (as well as transfusions of alcoholic beverages like wine and beer)—they were just plain weird.

On a related note about blood typing, since erythrocytes in people with type O blood have neither A nor B antigens, in theory, their blood isn't recognized as foreign by any recipient's immune system (no matter what the recipient's blood type may be). Because of this, those with type O blood are sometimes referred to as universal donors.

Likewise, since people with type AB blood (a blood group discovered by Landsteiner's colleagues several years later) have *both* types of antigens, they can theoretically receive blood from any donor. As a result, people with type AB blood are known as universal recipients.

Unfortunately, the labels *universal donor* and *universal recipient* are somewhat misleading since there are other antigens and antibodies in the blood besides those of the ABO system. In modern times, blood is carefully cross-matched and screened for pathogens and toxic substances before any transfusions take place.

The work of Vesalius, Al-Nafis, and Harvey contributed to the less-than-timely discrediting of Hypocrites' humoral system, but by the early twentieth century physicians and researchers knew that bacteria and other pathogenic organisms were the cause of most diseases. Drugs like aspirin appeared in the late nineteenth century and the first antibiotics followed some thirty years later. These new treatments rapidly replaced bloodletting as methods to combat many ailments and to reduce the discomfort from wounds, inflammation, and fever.

Surprisingly, though (or not surprisingly, considering just how long the procedure was employed), therapeutic phlebotomy has shown a positive effect in alleviating some symptoms—especially those tied to elevated blood pressure or increased blood volume.

For example, aneurysms occur when a weakened portion of an artery bulges outward like a blood-filled water balloon. As the heart beats, the aneurysm pulsates (like rhythmically squeezing a baseball-sized water balloon in your hand). In some instances, there is a warning—as the vessel walls stretch, pain receptors connected to the surface of the vessel are stimulated. Unfortunately, in most instances the aneurysm is painless and goes undetected. The real danger from an aneurysm is fairly obvious, and if this "water balloon" pops, the results can be deadly. Many strokes result from ruptured aneurysms in the brain and a burst aortic aneurysm can lead to massive hemorrhaging and death within a matter of minutes.

Aneurysms can occur for any number of reasons including high blood pressure and arteriosclerosis. In the latter, this "hardening of the arteries" reduces blood vessel elasticity and produces weakened regions of the vessel wall that can bulge under the pressure of the blood carried within it.

In the days before the discovery of penicillin, aortic aneurysms were a common side effect of syphilis. Regular bleeding of patients with syphilis-related aneurysms was a way of reducing their blood pressure and thus lessening the chance of having their aneurysms rupture.*

Bloodletting was also used to reduce the pain of angina pectoris (literally, "strangled chest"). Angina results from an insufficient supply of blood to the heart musculature (usually because of a blockage or constriction of the coronary arteries). Like the pain from a pulsing aneurysm, angina pectoris is a symptom rather than a disease. It's part of the body's warning system—telling it in no uncertain terms that something is *seriously* wrong. Today, vasodilators like nitrates are used to treat angina. They work primarily by increasing blood flow to the farthest reaches of the body (peripheral blood flow), which reduces blood pressure. Studies on the action of these drugs performed in 1970 indicated that therapeutic phlebotomy also lowered pressure within the resting heart.

Similarly, congestive heart disease is commonly characterized by an increase in blood volume, and the diseased heart has a tough time pumping this blood to the body. In modern times, drugs like diuretics are used to decrease blood volume (basically, more urine results in less plasma). But up until the 1960s, a viable means to achieve the same result was to perform periodic bleeding of the patient. Thankfully, as in the current treatment of aneurysms and angina pectoris, drugs have replaced bloodletting and there are few patients (or physicians for that matter) who would complain.

But before we get too carried away with the wonders of drug therapy it should be mentioned that there are several conditions that *still* call for bloodletting.

*In his book *Blood: An Epic History of Medicine and Commerce,* Douglas Starr hypothesized that early transfusion recipient Antoine Mauroy actually suffered from an advanced stage of syphilis (which is caused by the bacterium *Traponema pallidum*). Starr suggested that Mauroy's early positive results might have occurred after a transfusion-induced fever killed off the heat-sensitive pathogens for a while.

Porphyria (from the Greek word for "purple") is a disease of the blood that results from the faulty production of hemoglobin, which leads to the accumulation of red and purple pigments called porphyrins. For several reasons, it was referred to as vampire disease. This is because in one form (porphyria cutanea tardia), porphyrins concentrate in the skin. When exposed to sunlight, these abnormal pigments cause damage to surrounding tissues, leading to symptoms like severe skin rashes and blistering. These light-activated toxins and the disfiguring damage they cause may have been a source for tales of vampyres and their destruction by sunlight. Additionally, in some forms of porphyria, severe anemia can result in a pale, spectral appearance, and as the gums recede (from lack of oxygen and nutrients), the teeth become more exposed, causing the canines to take on a fanglike appearance.

In the acute form of porphyria, exposure to certain substances (e.g., alcohol) can trigger severe neurological disturbances—a fact that now appears to explain the "madness" of England's King George III (the king mentioned in the Declaration of Independence). Starting in 1762, the British monarch began to have bouts of serious health problems. These often began with flulike symptoms, but they soon segued into depression and strange behavior—racing with a horse, pretending to play a violin, and claims that he could raise the dead. In another instance, George III was reported to have talked continuously for twenty-six hours, his speech garbled and repetitive.

The king's medical treatment was remarkably similar to that afforded (or inflicted upon) his American contemporary, George Washington. He was bled, blistered, purged from both ends, and cupped.* Leeches were also used to draw off presumed excesses of blood. Sometimes the king went years without an attack, but

*In this technique, a glass cup was inverted over a flame, heating the air within. The cup was then placed flush against the patient's skin, and when the air within it cooled, a vacuum formed, which was believed to draw out toxins from the body.

invariably they returned, and in 1811, George III's doctors certified him to be permanently disabled.

For many years, researchers were puzzled by the sudden onset of bizarre behavior and strange outbursts that began to affect the king, especially after he'd reached his fifties. In the 1960s, two authors used information from the king's medical records (including the observation that the monarch's urine was the color of port wine) to conclude that George III had suffered from porphyria.

Recently, additional evidence concerning the king's affliction has come to light. The examination of several strands of King George's hair revealed arsenic levels that were three hundred times the toxic level. Professor Martin Warren combined this new information with data from the earlier study to conclude that "the madness of King George" was brought on by porphyria but that the monarch's bizarre behavior was apparently triggered by arsenic contained in the very medications his doctors had prescribed to treat his madness.

What's the connection between porphyria and bloodletting? Although drug treatments now exist to alleviate the symptoms of this incurable illness, therapeutic phlebotomy is still employed to reduce blood volume—which decreases the levels of porphyrins in the plasma. This ancient technique has been effective in providing relief from the painful and often mentally debilitating symptoms of porphyria. In fact, therapeutic phlebotomy may have been responsible for George III's ability to undergo dramatic recoveries from his illness—until his arsenic-laced medications kicked in, that is.

There are several other conditions where bloodletting is still used as a therapeutic measure to reduce iron levels in the blood.

Recently, researchers have found evidence that patients suffering from the viral disease hepatitis C respond better to treatment with interferon if they are first bled to induce a mild state of iron deficiency. Interferons are small proteins naturally produced by cells like macrophages in response to a viral attack on the body. The interferons diffuse into uninfected cells where they act as a shield, interfering with the virus's ability to enter and infect the cell. Since viruses can replicate only within cells whose reproductive machinery they have hijacked, preventing their entry into uninfected cells is a vital part of the body's antiviral defense. Interferons also attract and stimulate killer T cells that attack cells infected with viruses. Scientists are still trying to determine why lowering plasma iron concentrations through therapeutic phlebotomy increases the efficiency of interferons (which are used to treat diseases ranging from hepatitis B and C to certain types of leukemia and genital warts).

Bloodletting is also used to treat a specific form of diabetes. Insulin, a hormone secreted in the pancreas, helps to adjust blood glucose levels by enhancing its absorption and use by the body. High plasma concentrations of the iron-containing protein ferritin may damage the cells that secrete or respond to insulin. This leads to a condition called high-ferritin type 2 diabetes (in which elevated iron levels ultimately result in dangerously high blood glucose levels). Studies have shown that insulin resistance in patients with high-ferritin type 2 diabetes improved after three bouts of bloodletting (in which five hundred milliliters of blood was drained every two weeks for a total of six weeks).

Other conditions still treated by therapeutic phlebotomy include hemochromatosis, in which the digestive tract absorbs too much iron (resulting in an overload of tissue-damaging iron deposited in places like the liver and pancreas), and polycythemia, in which blood cells are produced at an uncontrollable rate. In many instances the only way to treat these afflictions is to reduce the amount of blood.

Ultimately, then, there are several relatively rare disorders that appear to benefit from treatments involving therapeutic phlebotomy. But when comparing these instances to the fact that historically, bloodletting was prescribed for basically everything, it's easy to see that although the practice isn't quite dead it is certainly a medical relic.

There is, however, a form of medicinal bloodletting that has stood the test of time. Unlike the techniques described thus far, this one utilizes an ancient worm that has been securing blood meals for far longer than its parchment-winged counterparts. Like vampire bats, these creatures do their bloodletting far more efficiently than any technique cobbled together by humans—and recognizing this, *Hirudo* has been employed medicinally since the time of the pyramids. Nearly abandoned as a therapeutic tool in the twentieth century, the leech has undergone career rejuvenation as modern surgeons have recently turned to their ancient ally.

A skillful leech is better far, than half a hundred men of war.

—Samuel Butler

Rulers who neither see, nor feel, nor know,
But leech-like to their fainting country cling,
Till they drop, blind in blood, without a blow.

—Mary Shelley

6.

A BEAUTIFUL FRIENDSHIP

On the Ulanga River, East Africa
At the start of World War I

As Humphrey Bogart's character Charlie Alnutt struggles to haul the *African Queen* through the reedy tributary that he *hopes* will lead to Lake Tanganyika, he is utterly unaware (and just as unconcerned) that his movements are producing waves of pressure change in the chest-high water. The waves carry away from his body like the ripples radiating from a pebble someone might have flipped into a goldfish pond.

Charlie shoots a quick glance over his shoulder. Rosie (Katharine Hepburn) looks worried. "Don't worry, old girl," he says with as much cheer as he can muster.

Rosie's angular face breaks into a smile, but her body language betrays her. "I'm just fine, Mr. Alnutt," she says. Her New England accent has a distinctive quaver, but now Alnutt is alarmed at something else in her voice—something even worse than fear. This is more like doubt.

"Mr. Alnutt, please take care down there."

"Sure, Rosie, sure," he says, giving her a wave. Only after turning away does he snarl at the recurring image of the skinny missionary dame pouring out *all* of his gin.

Crazy old broad, he thinks, moving forward once again and wincing as the towrope bites into his shoulder.

Below the surface of the dark water, Alnutt's movements have sounded an ancient alarm, and as they have done since before the age of dinosaurs, the prehistoric hunters react instinctively.

Several resident leeches have been clinging to the reed shafts, which are to them the size of good-sized tree trunks. They hold on using a suction cup located near their tail end, and so their bodies hang free, drifting in the easy current. Combined with their green coloration, this mimicry gives them the appearance of reeds and serves to protect them from the hungry fish and birds that patrol the swamp.

Mostly the leeches just wait. Time means nothing to them, but with the sounding of a silent, internal alarm, they are instantly driven by a single, mindless directive.

To a human it can roughly be interpreted as FOOD.

Over millions of years, natural selection has equipped these leeches with an array of adaptations that make them well suited for a blood-feeding lifestyle. They include several sensory systems that provide a steady stream of information about their environment and the potential predators and prey that move through it. Ten eyespots are arranged in rows near the leech's head end, but unlike vertebrate eyes, with their complex focusing ability, these photosensors are specialized at detecting movement and sudden changes in light intensity.

At this distance, between the reeds and Charlie's boat, the eyespots register nothing out of the ordinary. Instead, each of the leeches inhabiting this small section of swamp feels a slight sensation run down one side of its body. Holding on with their caudal suckers, they extend themselves to full length, then freeze into an alert posture. The leeches hold this position for several seconds, and as they do, their sensory systems become attuned for additional incoming stimuli.

There. The vibration has come again.

SAME SIDE

STRONGER

The leeches briefly bend their bodies into horseshoe shapes, a position that enables them to store the potential energy that will be released an instant later when they spring away from the reeds that have held them above the mud.

Their simple brains are unaware that their bodies are covered by hundreds of tiny touch detectors. Nor do the leeches perceive that some of these mechanoreceptors* have been activated by the first wave of pressurized water radiating from Charlie Alnutt as he tows the *African Queen* into their lair.

Now the free-floating leeches hesitate for a fraction of a second until another wave of disturbed water hits them. Instantly, they are moving in choreographed unison, adjusting their direction so that the waves of sensation become stronger and stronger, hitting them at their head ends rather than their sides.

Without thought, the hunters begin undulating their bodies faster and faster, converging on the food source from five different directions, like swimmers from some unreleased Esther Williams horror film.

*Mechanoreceptors are specialized sensory structures that are stimulated by physical contact (just like chemoreceptors are stimulated by chemicals).

Five feet away.

Three.

There are flashes of movement. An enormous dark shape—and suddenly the leeches are struggling against tidal waves that batter them from all sides.

Although they cannot register awe (or any other emotion for that matter), this is the largest food carrier that the leeches have ever tracked.

One foot away.

The chemical signals grow stronger and the leeches can feel the heat from the food as it moves in a pulsating torrent—close now.

Inches away.

One of them batters up against a gigantic moving wall.

FOOD

As the leech tries desperately to secure its anterior suction cup, a tremendous wave smashes it away. The creature cartwheels through the turbulent foam, spinning wildly until the jagged tip of a broken reed spears it.

The skewered leech is thrown sideways. Its muscular contractions are still strong, but now they send the creature into an uncontrolled downward spiral.

Five other leeches have followed a concentration gradient of chemical cues, finally locating a breach in the impregnable, woven wall. They swarm in, fanning out and using their suckers to secure purchase.

In the mud below, another hunter stirs. The twelve hundred facets that make up the dragonfly larva's compound eyes register movement as the wounded leech corkscrews toward it—slower now—a top running out of spin.

The predator hurtles upward, capturing the broken reed with six powerful legs. The larva's mouthparts sink into the leech's body, working with robotic precision to free the meal from its skewer.

From a gap in the reeds, a two-foot-long perch watches with growing interest and a hunger of its own.

Inside the impregnable woven wall, the leeches use three sets of chitenous jaws to saw their way through the thin, wire-covered rind that separates them from the food. Waves of peristaltic muscular contraction run down their elongated bodies and within seconds they begin to—

Five tiny lives blink out in succession. There is no real pain. Just the sensation of feeding . . . and then . . . *nothing.*

On the deck of the *African Queen,* Rosie helps Charlie brush off the last of the dead leeches and the salt she's thrown on them.

Charlie Alnutt shudders. "If there's anything in the world I hate, it's leeches—filthy little devils!"

Long Island, New York
September 2006

As I stepped out of my car, Albert Einstein's twin brother gave me a friendly wave. "Rudy Rosenberg," the man said, extending a hand.

He wore a red-and-white pin-striped shirt and a blue bow tie. His white hair fell in longish corkscrews around his face. I guessed that he was in his midseventies.

"Let's go in the back way," Rudy suggested, stopping at a keypad before punching in the combination. I followed him into a rather plain-looking industrial building around the corner from the Long Island Expressway. I was a bit surprised that the sign out front read "Accurate" and not "Leeches USA." I guess I'd been expecting something flashier—a stream of blood or at least some suction cups.

"There are three companies housed here," Rudy told me several minutes later as I settled into a seat in his office. The room was comfortably cluttered—every inch of space crammed with artwork, books, journals, and memorabilia. "Accurate Chemical and Scientific, Accurate Surgical and Scientific Instruments, and Leeches USA."

I perked up.

He looked at me over the top of his wire-frame glasses. "It's always the leeches that people are interested in."

Rosenberg, a Holocaust survivor, began by explaining that leeches were one of the oldest therapeutic techniques known to man.

"They've been identified from Egyptian hieroglyphics, thirty-five hundred years old. In Sanskrit writings from one thousand BC, leeches were described as treatments for snakebites. The Greeks used them for headaches."

I knew from my investigation on bloodletting that in the second century Galen had prescribed leeches as a means to reduce "plethoras"—perceived excesses in the volume of blood (one of the four humors)—and that the word *leech* was actually derived from *loece* or *laece,* an old Anglo-Saxon word meaning "to heal."

"And for all our technology and a multibillion-dollar drug industry, leeches are still being used in hospitals today. In fact, in 2004 the FDA approved *Hirudo medicinalis* as a medical device."

I wondered if I'd heard him correctly. "Device?"

"That's right. The medicinal leech is only the second-living creature to ever receive that designation."

Rudy knew the answer to my next question, so before I could ask it, he said, "Maggots."

"Maggots?" I parroted.

"Sure. Doctors figured out hundreds of years ago that if they applied maggots to an open wound, they'd only eat the dead or decaying tissue—not the viable stuff."

"Cool," I replied, ever the scientist. "So what about leeches?"

"Most of ours go to hospitals or medical centers where doctors perform reattachment surgeries—ears, scalps, noses, breast reconstructions too. Somebody walks into a hospital carrying their finger in a bag of ice—we get a call."

Rudy explained that during delicate procedures like the reattachment of an ear or finger, tiny arteries and veins are reconnected with microscopically thin sutures. With their thicker, sturdier walls, arteries are easier to work on, but the thin-walled veins that return blood to the heart are often problematic. As a result, blood pumps freely into the reattached tissue (via the surgically mended arteries) but it often pools up owing to the compromised venous flow out of the area. If not corrected, this "venous congestion" leads to blocked circulation and, ultimately, the death of the reattached tissue, which starves from the lack of oxygen and nutrients.

"At some point, someone figured out that by placing leeches near the site of the reattached tissue you could establish a sort of artificial circulation."

"How's that?"

"The leech drains the accumulating blood so that new arterial blood can arrive on the scene. This fresh blood brings with it the ingredients necessary to repair the damaged blood vessels and stimulate the growth of new veins into the area."

"Got it," I replied.

"They might use ten to twelve leeches on a finger reattachment and several hundred after a scalp reattachment surgery."

Rudy explained how over a thousand leeches had been used to save the remaining portion of one Canadian man's leg after cancer had forced surgeons to perform a partial amputation.

"Still, it wasn't until the 1980s that leech use progressed from being a last resort to becoming a rather standard procedure."

"Why the reluctance?" I asked.

Rudy leaned across his desk and lowered his voice. "Most surgeons thought that it reflected badly on them if they needed to use leeches. Initially, it was quite a tough sell. Now we ship thousands of them each year—all over the U.S. and Canada."

I asked Rudy if all of the leeches he sold were headed for reattachment duty.

"Mostly," he replied, "but veterinarians are using them, too, for reattachments in dogs and sometimes to treat swollen ankles in horses. They're also popular in teaching labs." Rudy explained how leeches had relatively huge neurons (nerve cells), making it easier for students to study nervous system function at a cellular level.

The leech-meister could tell that I was impressed but he wasn't quite finished.

"At Beth Israel Medical Center, studies are showing that leeches can be an effective treatment for pain and inflammation, especially for people suffering from osteoarthritis. They're one of the first hospitals in the country to offer this type of leech therapy."

Rudy sat back in his chair and grinned like a proud father. "Quite an accomplishment for a creature that most people consider a repulsive, bloodsucking worm."

I nodded, as if I'd been a leech fan all my life. I had to admit (although not to Rudy), that even though I studied vampire bats for a living, leeches had *always* given me the creeps. In fact they were right up there with clowns and televangelists. Now, sitting in front of this enthusiastic cheerleader for Team *Hirudo,* I was starting to see his tiny co-workers in a completely different light.

Leeches belong to the phylum Annelida (the segmented worms). The group contains around twelve thousand species and has a worldwide distribution. In addition to leeches (which belong to the class Hirudinea), annelids also include the earthworms (class Oligochaeta) and their freshwater relatives, and as well as marine wigglers like sandworms and bloodworms (class Polychaeta).

Numbering around 650 species,* leeches can be found in cool freshwater environments or stagnant tropical mud wallows. About 20 percent of leech species are marine and their habitats range from shallow coastal waters to thermal vents located twenty-five hundred meters below the surface. Other leeches are completely terrestrial and they too exist in a diverse array of habitats: from tropical rain forests to the sub-Himalayan hills of northern India. There's even an unpigmented species that inhabits a single cave in New Guinea where it feeds on the blood of bats. Although leeches are most famous (or infamous) for their bloodsucking ability, many species are predatory rather than parasitic. A few species even provide some benefit to their rather oblivious hosts.

Annelids range in size from less than a millimeter in length to over three meters long, in the case of a giant Australian earthworm. They're called segmented worms because their bodies are composed of ring-shaped segments (annuli) stacked one upon another (kind of like the Michelin Man—but with no arms or legs and a bit more slime).

From an evolutionary perspective, the adaptive advantage of segmentation (also referred to as metamerism) appears to be related to the fact that annelid bodies are literally divided into a series of more or less independent sections (walled off from each other by thin septa). In the distant past, these annuli may have

*To put the diversity of the Hirudinea and its 650 members into perspective, there are roughly five thousand species in the class Mammalia and about thirty thousand species of bony fishes in the class Osteichthyes. Whenever we vertebrate biologists get too pumped up over the vast number of animals we have to study, it's often sobering to check out the invertebrate class Insecta. This group wins the Animal Diversity Contest hands-down, with estimates of well over one million living species, including *over three hundred thousand species of beetles*!

served as an early framework for regional specialization of the animal body into a head, midsection, and tail end.

Apparently, segmentation also allowed annelids to move more efficiently than their nonsegmented ancestors. This is because along with a metameric body plan, another adaptation evolved in this group—a body cavity known as a coelom. This fluid-filled chamber is part of the worm's hydrostatic skeleton, so that when muscles that encircle each body segment contract, the compressed fluid within the coelom is forced toward the head end, projecting the front of the body forward. You can demonstrate this type of locomotion for yourself by squeezing one end of an elongated, water-filled balloon around the middle. Your hand represents the annelid's circular body muscles, while the water and the expandable balloon interior act as the coelomic fluid and the coelom, respectively. In earthworms, as the body extends forward, the worm secures itself to the substrate with ventrally located pairs of microscopic, toothlike setae (or chaetae). The back end of the body is then pulled forward as longitudinal muscles that run down the length of its body are contracted. The vermiform crawling that results from this anatomical arrangement is quite a bit more efficient than the wild whipping and thrashing movements that characterize other worms like nematodes (commonly known as roundworms).

Leeches, however, are capable of employing an alternative form of locomotion, and in these instances they resemble inchworms (which are insect larvae, not real worms). This type of locomotion is possible because of the presence of a pair of sucker disks—one located near the head end (the anterior sucker) and the other near the tip of the tail (the caudal sucker).*

During inchworm crawling, the leech attaches its caudal sucker

*One Trinidadian genus *(Lumbricobdella)* has reverted to the burrowing lifestyle of its ancestors. Not surprisingly, this leech has lost its caudal sucker and moves through the soft ground much as an earthworm does.

to the substrate (which may be oriented horizontally or vertically). Next, the muscles encircling the body contract, extending the anterior end forward (as in vermiform locomotion). The anterior sucker then takes hold of the substrate and the caudal sucker releases its grip. Finally, the tail end of the body swings forward, planting the caudal sucker directly behind the anterior one. Inchworm crawling can be employed underwater or as aquatic leeches leave the water to lay their eggs. It provides an additional advantage by allowing the leech to move efficiently across vertical or smooth wet surfaces. Terrestrial leeches also use inchworm crawling to zero in on their warm-blooded targets. Potential meals can be tracked from a distance of about two meters away. This is done primarily through detection of vibrations produced as the prey (or host) moves through its environment as well as the carbon dioxide it exhales. Vision is also employed as photoreceptors provide the leech with information on changes in light intensity (e.g., from shadows passing nearby).

Many other leeches are quite adept at swimming, but unlike fish, which move by undulating from side to side, leech bodies bend with an up-and-down motion that is reminiscent of dolphins and whales. Leonardo Da Vinci (1452–1519) may have been the first to study leech locomotion and he actually spent quite a bit of time getting the intricacies of their dorsoventral undulations just right.

Unlike the parasitic torpedo attack launched against Bogart's character in the 1951 classic *The African Queen* (those "leeches" were actually made out of rubber), there are aquatic leeches that gain access to their hosts via different routes. Some hop aboard as their victims dip their heads for a drink. Entering through the nostrils, the leeches attach themselves to the mucous membranes that line their host's nasal cavity. There, in the warm, humid chamber, they feed and mature, safe from detection—at least for a while.

One famous story recounts how this type of leech attack

afflicted a group of Napoleonic soldiers, crossing from Egypt to Syria in 1799. As with any army in a foreign land, obtaining water was always a major concern and things were even dicier in the days before purification techniques (like boiling or adding iodine or chlorine) could make most water at least semisafe to drink. Apparently, some of these men drank water from a lake infested with tiny larval leeches. Unbeknownst to their hosts, the creatures quickly attached themselves and began to feed. Days later the men began to take ill and medical personnel were horrified to find their patients' noses, mouths, and throats carpeted by blood-engorged leeches. The doctors tore frantically at the vampires and it's not hard to envision a gory scene made all the more horrible by the cries of fearful, frantic men.

As in other similarly large groups of animals, there is a wide degree of variation in leech diets. About three-fourths of all known species are bloodsucking parasites, feeding primarily on the blood of vertebrates (including sharks, bony fishes, frogs, turtles, snakes, crocodiles, birds, and mammals). Parasitic leeches do not generally specialize on any one particular prey. For example, the medicinal leech *Hirudo medicinalis* usually feeds on frogs, but it will readily take human blood. And while we're on the topic, it should also be noted that dietary relationships between leeches and other organisms aren't all stacked in the leech's favor. Leeches are commonly fed upon by fish, birds, salamanders, snakes, and even other leeches.

According to Dr. Mark Siddall, a leech expert and curator at the American Museum of Natural History, the first leeches were freshwater blood feeders, but alternative feeding modes have evolved at least six times in various leech groups. Several of these lineages have become predators, feeding solely on invertebrates (like their earthworm cousins, snails, or even other leeches). Unlike parasitic leeches, which can survive for extended periods of time without feeding, predaceous leeches feed frequently (generally, every one to three days). Another difference is that the predators digest their food rapidly, while their parasitic relatives are able to hold blood within their guts in an undigested state for up to several months. Predatory species are also equipped with a highly mobile, hoselike proboscis that is first employed to probe the potential prey. Once this tactile reconnaissance is complete, the leech inserts the proboscis into the flesh of its soon-to-be meal and proceeds to suction feed for up to several hours—vacuuming out virtually all of the prey's soft-bodied interior. Often, the predator is joined by others of its kind, and the scene comes to resemble a Chinese buffet table attacked by senior citizens. In each instance, the ravenous feeding frenzy leaves behind only carnage.

Some leeches are neither parasites nor predators. Among these are several members of the family Branchiobdellidae. These species are remarkable for the manner in which they partition their highly specific habitat—the shells of the freshwater crayfish. Although the hard outer covering of these minilobsters might seem an unlikely place for leeches to hang out, there can be up to seven species living on a single individual. In what stands as an extreme example of microhabitat partitioning, different leech species inhabit different regions of the crayfish's body. For example, one type of leech lives attached to the crayfish's antennae, while another lives out its life adhering to the pincer-tipped legs (the chelipeds). Unlike parasitic leeches that feed on the blood of their hosts, the behavior of these branchiobdellids is more comparable to the previously mentioned endosymbiotic bacteria living in the guts of mammals like cows. Although not as important as bacterial endosymbionts are to their hosts, these *ecto*symbionts provide a service to the crayfish by grazing on the micofauna (algae, diatoms, and bacteria) that attach to its body.

This leech/crayfish association is not always positive for the host since some branchiobdellids don't provide the aforementioned cleaning services. A few species cause no harm—they're merely opportunists, consuming scraps of organic matter scattered about by the messy crayfish as it shreds its food into bite-sized bits. Another branchiobdellid, however, *is* parasitic. It lives within the crayfish's gill chamber where it feeds on the gill filaments and blood of the same freshwater crayfish whose bodies are so precisely partitioned by its nonparasitic cousins. Even worse, however, are the instances in which crayfish are infested by lethally high numbers of leeches, which can sometimes cover their bodies like living carpets, ultimately killing them.*

*Those readers interested in learning more about leeches (a lot more, actually) should consult Roy T. Sawyer's 1986 three-volume magnum opus *Leech Biology and Behavior.*

Early use of leeches by man reflected the importance of bloodletting as a therapeutic tool. Leeches also gave practitioners an alternative when "breathing a vein" wasn't appropriate. For example, leeches could be applied to parts of the body that were difficult or impossible to bleed by lancing or other means. Inflamed tonsils might call for leeches to be attached to the back of a patient's throat or the bloodsuckers might be employed to drain persistent hemorrhoids. Leeches were applied to the scrotum to treat the swollen testicles that resulted from gonorrhea and they were also commonly used to treat maladies of the female reproductive system. Additionally, leeches were the preferred method of bleeding women and children "who required a gentle withdrawal of blood."

In what is arguably the strangest use of leeches on record, the sixteenth-century French historian Pierre de Brantôme recounted how leeches were inserted into the vaginas of women on their wedding nights so that they could "seem like the virgins and maidens they used to be . . . so as the gallant husband who comes on his wedding night to assault them, bursts their bulb from where the blood flows."

According to Brantôme, battering this bogus maidenhead (or having it battered for you) invariably led to an annelid-assisted version of postcoital bliss: "And both are bloody and (there is) a great joy for both and in this way the honor of the citadel is safe."

Right.

Medicinal leech use reached its zenith in Europe in the first half of the nineteenth century, where Napoléon's chief surgeon, François-Joseph Broussais, ascribed to the idea that all diseases were a result of too much blood (Galen's plethoras again—fifteen hundred years later). As a result, Broussais prescribed leeches (and the always-popular "starvation") in much the same way that a physician today might recommend aspirin and bed rest. Given Broussais' tremendous influence on European medicine, the use of leeches exploded in the 1830s, with over forty-one million used in 1833. French troops were bled for every conceivable ailment. Some of them were treated with as many as fifty leeches at a time—so many, in fact, that they were said to be wearing glistening "coats of mail." Fashion-conscious ladies of the time even wore dresses "à la Broussais," decorating them with imitation leeches. Leech use was so heavy that the medicinal leech was driven to the point of extinction (and it remains endangered today). Eventually, they had to be imported from places like Asia.*

Given the degree of crushing poverty that existed at the time, leech collection became a steady, if not particularly pleasant, way to generate income. Nearly all that was required to start a dynamic career harvesting *Hirudo* was access to a lake, pond, or swamp that had leeches living in it. Leech collectors simply rolled up their pants (or skirts) and waded into the nearest leech-infested body of water. Then they stood around (swamps suddenly became *the* place to hang out) until a hungry leech or two swam by and decided to latch on to a leg or foot. Once the parasite had secured itself, the "lucky" collector would gently pull the leech off and place it into a basket. Presumably, those folks with

*According to American Museum of Natural History leech expert Mark Siddall, the leeches being cultivated today at places like Leeches USA aren't really *Hirudo medicinalis* but *Hirudo verbana* (a species that isn't protected by the Convention on International Trade in Endangered Species or approved for use as a medical instrument by the U.S. Food and Drug Administration). Just as important, it appears that wild leeches, from across Europe, comprise three separate species, creating at least the potential for three times as many anticlotting compounds. Rudy Rosenberg said that if the new classification is accepted he will petition to extend its approval to *Hirudo verbana*.

more time on the job were able to remove the leech after it had attached itself but before it had initiated a bite. For many, it appears that collecting leeches in this manner was far from a pleasant way to make a living. As described by the Reverend J. G. Wood in 1885:

> The Leech-gatherers take them in various ways. The simplest and most successful method is to wade into the water and pick off the leeches as fast as they settle on the bare legs. This plan, however, is by no means calculated to improve the health of the Leech-gatherer, who becomes thin, pale, and almost specter-like, from the constant drain of blood, and seems to be a fit companion for the old worn-out horses and cattle that are occasionally driven into the Leech-ponds in order to feed these blood thirsty annelids.

"Today, leeches are raised in tanks partially filled with distilled water and refrigerated," Rudy said.* "They're hermaphrodites, so everybody gets pregnant. After *Hirudo* mates, it crawls out onto land to lay its cocoons—which look like foamy little footballs. The babies hatch from the cocoons in around three weeks."

"Do you raise your own leeches here?" I asked.

"No. We get them as adults from a company called Ricarimpex. They've been in business since 1845."

"Yikes," I said, impressed that any company could stick around that long—no less one that sold leeches and whose name seemed to have been thought up during a night of drunken Scrabble. I even briefly thought about suggesting a motto ("With a name like Ricarimpex you *know* our product sucks!"), but I held back.

"There's another company, Biopharm, that was started by Roy Sawyer in South Wales."

"Biopharm? Cool name," I said, a bit too quickly and Rudy shot me a puzzled look.

Back in the early nineteenth century, several companies, including Ricarimpex, sprang up in the marshy regions near Bordeaux, France. (The former company name was actually Ricard-Debest-Bechade.) These leech suppliers flourished because of the sudden demand that resulted from Dr. Broussais' "if it's sick, stick a leech on it" brand of medicine. Leeches were now required in vast numbers and the old capture technique of "wade in and wait" was soon replaced by leech husbandry and the establishment of permanent breeding pools. New roads and rail links meant that leeches could be sent farther and farther afield. The

*Like many freshwater leeches, *Hirudo* is most active in cold water (around 42–45°F, or 5–7°C). Presumably, this is when their aquatic prey are sluggish or inactive and therefore easily attacked. Many leech species become stressed out in warm water, probably owing to a decrease in dissolved oxygen. In some freshwater species that prey on fish, a rapid increase in water temperature coincides with a detachment from their prey, whereupon the leeches reproduce and die.

practice of leeching spread to the United States, although American species were found to be deficient because of their relatively puny size.

Most, but not all, leech use by humans was related to bloodletting. In 1850 Dr. George Merryweather came up with a rather remarkable, nonmedicinal use for leeches. He did so after pondering a short section from Edward Jenner's poem, "Signs of Rain":

The leech disturbed is newly risen;
Quite to the summit of his prison.

Merryweather interpreted this line as a reference to the medicinal leeches' sensitivity and response to the electrical conditions in the atmosphere, and he sought to use this to predict upcoming

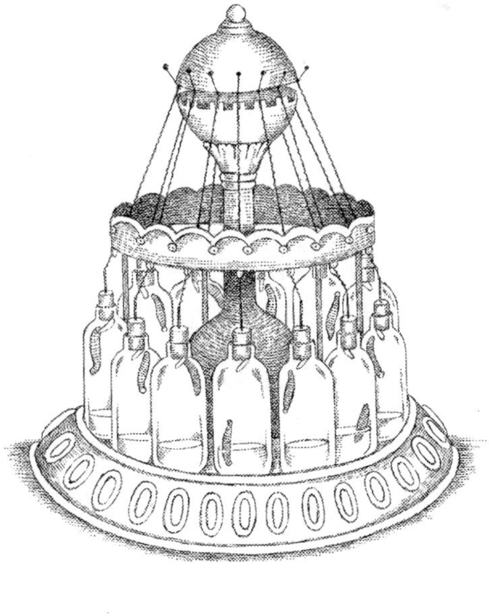

storms. The instrument he created was the Tempest Prognosticator, and it consisted of twelve pint bottles, containing about an inch and a half of rainwater. The bottles were set in a circle beneath a large bell. This arrangement, Merryweather stated, allowed the leeches to be within sight of one another, thus preventing them from "feeling the affliction of solitary confinement." At the top of each bottle was a narrow metal tube and in each tube a tiny piece of carved "whalebone" attached to a wire.* The twelve wires led up to the bell where they ended in miniature hammers. The contraption was designed so that the whalebone would be dislodged if a leech entered the tube—something it would do only when bad weather approached. A shift in the position of the whalebone pulled the wire, causing the bell to be struck by the hammer. The more leeches that rose, the more times the bell was struck, thus indicating the relative strength of the approaching storm. Although the Tempest Prognosticator functioned successfully, scientists actually think that the leeches were responding to changes in barometric pressure rather than sensing electrical activity.†

Merryweather proudly displayed the instrument (also called a Leech Barometer) in the 1851 Great Exhibition at the Crystal

*Whalebone is the layman's term for baleen, which isn't bone at all. Baleen is composed of the waterproof protein keratin and grows in plates in the mouths of filter-feeding whales (like the blue whale, *Balaenoptera musculus*). Biochemically identical to hair, baleen tends to curl or frizz in humid air, while straightening out in dry air. This property may or may not explain why Merryweather used it in his contraption—since the definitive function of this substance remains uncertain from the somewhat vague descriptions.

†This behavior is similar to that reported in a study of juvenile blacktipped sharks *(Carcharhinus limbatus)*, schools of which were reported to respond to barometric pressure changes associated with approaching tropical storms by moving to deeper water. Darrin Lunde, of the American Museum of Natural History, suggested that the vertical migration of Merryweather's leeches might be explained by the fact that aquatic leeches move out of the water and onto land only when conditions are humid and wet—as they generally are after significant drops in barometric pressure.

Palace in London. He implored government officials to utilize his design, envisioning a protective shield of bell-ringing leeches encircling England's coastline. He also lobbied that his Tempest Prognosticators should be placed aboard every ship in his country's great fleet. Instead, the Royal Navy opted for an annelid-free barometer (the storm glass) designed by Captain Robert Fitzroy. Fitzroy had used the device years earlier on the HMS *Beagle*—on a voyage that would become famous for entirely different reasons.

Aside from the unfortunate Dr. Merryweather, most people who came into contact with leeches did so because they were ill. A number of historical figures were bled by them—although none of them went on to become spokespersons for the treatment.

In April 1824 Lord Byron, who was on a military campaign in Greece, suffered a series of seizures, possibly related to the fact that he was addicted to drugs, had previously contracted both gonorrhea and malaria, and might also have had an eating disorder. Hospitalized and wracked with fever, the poet was disgusted to learn that his physicians had proposed attaching leeches to his brow to treat his elevated temperature.

"A damned set of butchers," Byron called them, between bouts of delirium and paranoia. He was *somehow* convinced that his doctors were going to kill him.

Byron's condition continued to deteriorate until, in a weakened state, he submitted to his doctors' recommendations. One could almost hear a collective sigh of relief from the physicians: the great poet had *finally* come to his senses.

The healers acted immediately, withholding water and attaching somewhere between twelve and twenty leeches to Lord Byron's fevered forehead. The hungry creatures did their job, reportedly draining off two pounds of Byron's blood. Unfortunately, the pathogen-packed poet died the following day. He was thirty-six years old.*

Leeches were also commonly used to treat strokes, and although there are discrepancies concerning Soviet strongman Joseph Stalin's last days, there is general agreement that he died on March 5, 1953, several days after being stricken by a massive cerebrovascular accident, the same ailment that had claimed FDR eight years earlier.

Summoned, some contend, up to thirteen hours after he was discovered in a puddle of his own urine, Stalin's terrified doctors† bled the dying dictator with a total of eight leeches, attaching them behind his ears. With trembling hands, the physicians sponged down "the Boss" with aromatic vinegar, then tried injecting him with camphor and caffeine (and, quite possibly, anything else they could find laying around the *dacha*). But the heroic efforts of Stalin's physicians were in vain. According to Stalin's daughter, Svetlana, the stricken despot reared up at the last moment, in what she interpreted as a final tirade against those present in the room. Then, Joseph Stalin dropped dead, urine soaked and oozing blood from behind his parasite-pierced pinna.

In the years since his death, there have been accusations that Stalin's medical treatment may have been deliberately delayed (although there was apparently a general consensus that thirteen

*Byron was long rumored to have been the true author of *The Vampyre,* a work credited to his former friend, the physician John Polidori. Evidently, Byron came up with the tale during the laudanum-fueled summer of 1816 as he and his friends (including Mary Wollstonecraft Shelley) spun tales of horror during their stay at Lake Geneva. After Byron discarded the idea, Polidori expanded it into a short story whose leading character, the aristocratic Lord Ruthven, would become an inspiration for Bram Stoker's Count Dracula.

†Stalin had recently initiated a purge of physicians (including his own) after claiming that they were part of a sinister Jewish plot to destroy the Russian people. The traitorous doctors, Stalin claimed, were murdering Russian statesmen. According to rumors, mass deportation of Jews to Siberia was to begin on March 5, 1953.

hours wasn't *really* that long of a delay). And, in any event, one could argue that given the state of Soviet medicine in 1953, the wait might not have been such a bad thing.

Finally, James Joyce, the author of *Ulysses,* underwent periodic treatment with leeches, and the results were only somewhat less grim than those experienced by Byron and Stalin. In addition to enduring eleven eye operations during the second half of his life, the Irish writer occasionally had leeches applied around "the outside of his eyes." It's not clear, however, if this refers to somewhere on his face (surrounding his eyes) or, less likely, one hopes, attached to the *surface* of the eyes themselves. Unfortunately, these treatments were unable to halt the progression of his glaucoma-induced blindness, although Joyce's death in 1941 was unrelated and resulted from a perforated ulcer.

One problem physicians encountered during leech use was that the leeches had an annoying tendency to wander off from where they'd been placed. In cases where leeches were inserted into a body orifice, a lasso of thread was first thrown around the creature to prevent it from taking an unexpected detour. Similarly, when leeches were applied just outside a body opening (as in the treatment of a boxer's cauliflower ear), a wad of cotton might be inserted into the nearby hole to keep the leech from venturing in. Presumably, this became a popular measure in the nineteenth century after a leech was reported to have ascended into one patient's uterus, while in another case, the fast-moving creature disappeared up someone's rectum.*

*I wondered if any such contrivances had prevented the "wedding night" leeches described by Brantôme from bailing out when attacked by what must have seemed like a blind but energetic relative.

Rudy Rosenberg explained that in addition to the ease of placing leeches in otherwise hard-to-bleed locations, the resulting bites are painless since one of the substances found in leech saliva is a sensation-numbing anesthetic.

"You might feel a slight pinch," Rudy Rosenberg told me. "But after that, nothing."

The ability to deliver bites without upsetting the potential meal is an important adaptation for a lifestyle in which the stealthiest individuals are the ones who survive to reproductive age. As with the bite of the vampire bat, if the leech's prey felt anything more than a minor annoyance, the chances for a successful feeding session would be greatly diminished.

"The medicinal leech has three jaws," Rudy continued. "They're arranged in a Mercedes-Benz cross and each jaw has around a hundred teeth.* Muscles that control a hood of skin at the head end of the leech form an airtight seal around the bite site. Then other muscles drive the jaws back and forth through the skin like a saw."

Leeches that specialize in larger hosts, such as water buffalo or even elephants, have larger jaws with even more denticles. This enables them to cut through the thick, tough hides that they encounter.

As mentioned earlier, the vast majority of leech species, including *Hirudo medicinalis,* have an additional suction cup near the tail.† This caudal sucker doesn't have denticles associated with it, though, since it functions primarily in locomotion and by providing an additional point of contact for the leech to attach itself to a host's body.

Sucker attachment actually has two components: adhesion via glandular secretions and the generation of negative pressure by a

*Leech teeth are more accurately referred to as denticles since they don't share an evolutionary origin with vertebrate teeth.

†This necessitates a dorsally located anus.

circular disk of muscle that leech expert Roy T. Sawyer described as "a finely tuned suctorial device." The inner surface of this sucker is lined with glands that secrete a mucuslike polysaccharide that aids in adhesion (in much the same manner that a wad of wet toilet paper clings to a ceiling of a dorm bathroom). In some species that prey on fishes, this adhesive substance is actually a flesh-dissolving enzyme that leaves permanent scarring at the attachment site (which is usually located near the base of a fin).

Rudy explained that none of *Hirudo medicinalis*'s adaptations would have elevated it much beyond the status of "rarely encountered but truly disgusting annoyance" were it not for the leech's ability to produce one of the most potent anticoagulants known to man. Identified in the late nineteenth century, the substance (now known as hirudin) was purified in the 1950s and cloned some thirty years later. Like the clot-preventing substances found in vampire bat saliva, hirudin has helped to transform the medicinal leech and some of its relatives into superbly evolved sanguivores. When released into a bite-inflicted wound, hirudin works in combination with a painkilling anesthetic, a blood vessel–widening vasodilator, and another substance (hyaluronidase) that promotes local spreading of the leech's potent salivary stew.

Initially, blood flows from the wound because of negative pressure produced by the anterior suction cup, but before long, suction is enhanced by wavelike peristaltic contractions of the leech's digestive tract. The leech remains attached to its host for up to an hour, sometimes ingesting ten times its body weight in blood (medicinal leeches consume about ten milliliters of blood per meal). Once satiated, the leech (bloated like some blood-filled, organic football) releases its suction grip and falls off. Depending on the species, leeches can survive up to three years between blood meals.

For the host, even after the leech drops off, the effects of the bite are far from over. Bleeding from the wound site continues for up to ten hours, and like the aftermath of a vampire bat attack, this can leave a considerable mess. Even under controlled conditions, patients undergoing leech therapy sometimes require transfusions to replace blood lost *after* the leech finishes feeding.

In addition to an often-significant loss of blood, there can be additional problems for the leech-bitten. *Aeromonas hydrophila* is an endosymbiotic bacterium that lives in the leech's gut. This organism can produce infection at the wound site as well as diarrhea for the human host. Because of this, leech treatment is often accompanied by a regimen of antibiotics.

What's *Aeromonas* doing inside the leech?

Researchers believe that in addition to assisting in digestion, *Aeromonas* produces metabolic by-products that serve as vitamins and essential amino acids for the leech. Like other endosymbionts, *Aeromonas* gets a safe place to live, feed, and reproduce. Similarly, other species of endosymbiotic bacteria inhabit the guts of mosquitoes and vampire bats.

Often, leeches carry even nastier pathogens in their saliva—including blood-borne parasites known as trypanosomes. These protists (notable for their prominent, whiplike flagella) are responsible for several serious diseases in their human and nonhuman hosts. Thankfully, leech-to-human transmission is rarely if ever reported—probably owing to the fact that leeches do not carry the infective stages of these flagellates—which are much more likely to be found in creatures like assassin bugs (a family of insects that have evolved a long and complex relationship with trypanosomes).

"Only *Hirudo medicinalis* has the right combination of traits," Rudy said, with a measure of pride. "Size, bite technique, and just the right amount of hirudin."

"How do you get them to bite where you want them to?" I asked.

"In the old days, they used a quill or some other hollow tube. They'd place the leech into the tube then hold it against the spot where they wanted it to bite."

I also learned that barber-surgeons or other "leechers" would place several *Hirudo* into a teacup, invert the cup, and hold it in place on their patient until the leeches got the message and latched on.

"In modern times, though, they'll just cut a small hole in a sterile gauze pad, place the hole over the designated spot, then stick the leech's head near the hole. The gauze keeps the leech from wandering off, but you've *still* got to keep an eye on them."

"Do they ever turn down a meal?" I asked.

"Oh, they can be *very* picky," Rudy said with a laugh. "Leeches hate perfumed soaps or smelly hair sprays. Often, they won't attach if the person is a heavy smoker or if they've eaten garlic recently."

I must have looked confused.

"The patient, that is."

"Got it," I said. "So how do you deal with a leech who plays 'hard to bite'?"

"Most of the time if you clean and shave the potential attachment site, then sprinkle on a couple of drops of blood, sugared water, or even milk—you can whet their appetite."

"And what if that doesn't work?"

"You can try abrading the skin a bit. Other times you can get leeches in the mood by immersing them briefly in diluted wine or warm, dark beer. And if they *still* won't bite, we just tell the physician to grab another leech."*

"What happens to the leech *after* a treatment?"

*In a 1994 study, Norwegians Anders Baerheim and Hogne Sandvik proved that medicinal leeches took twice as long to bite after they had been briefly submerged in Guinness stout (187 seconds to 92 seconds for the water control). According to the

Rudy frowned at this one. "Well, unfortunately, the treatment turns out to be their last meal," he replied. "Scientists have figured out that some human blood cells remain alive within a medicinal leech for up to six months." Evidently, they're still trying to figure out how this works but it appears to have something to do with the *Aeromonas* bacteria releasing chemicals that kill off other bacteria that might be present. Bacteria that would have presumably contributed to the destruction of the blood had they been alive and active.

"Anyway, blood-borne diseases can be transmitted if leeches get reapplied to someone else—so the bottom line is that once they feed, leeches are treated as medical waste."

"And how do you dispose of them?"

"Submerse them in alcohol, generally," Rudy said, then he looked at me over the top of his glasses again. "But you *never* want to flush them down the toilet."

I immediately took the bait. "Why's that?"

"Well, one day I got a call from a hospital. For some reason they'd decided to get rid of their leeches—so they flushed them."

"And?"

I noticed that the leech maven could barely contain his glee.

---

authors, "After exposure to beer some of the leeches changed behaviour, swaying their forebodies, losing grip, or falling on their backs." Presumably after the leeches sobered up, they were used to test an old wives' tale that a little soured cream applied to the skin would encourage leeches to feed more readily. The results of their study did not support the claim. Finally, Baerheim and Sandvik applied the leeches to arms that had been smeared with garlic. The researchers reported that the leeches "started to wriggle and crawl without assuming the sucking position" and "did not manage to coordinate the process" of biting. The experiment was halted after two leeches dropped dead within two and a half hours of exposure to the garlic-swabbed limb. Lest the reader question the seriousness and scientific relevance of this study, the authors, in their acknowledgments, thanked a local brewery "for supplying sufficient amounts of their precious liquid to satisfy the needs of all participants of the study." They also thanked the leeches for their enthusiasm, assuring readers that the worms were "by all accounts grateful to Hogne Sandvik for supplying his own precious liquid."

"Well, apparently the leeches loved the idea. They're escape artists to begin with—and pretty soon they started showing up in toilets all over the hospital. The hospital people were kind of frantic by the time they called me."

"What did you tell them?" I was already envisioning some truly unique target practice opportunities.

Rudy threw up his hands. "I told them to get a net."

In the late nineteenth century, medical advances began to overshadow ancient techniques like bloodletting and leeching. Physicians shifted their focus from humeral imbalances to bacteria as the causative agents for diseases and infections. A few daring researchers even claimed that bleeding a patient might do them more harm than good. Not surprisingly, the medical community showed some initial resistance to revamping, so to speak, what had been a longstanding therapeutic procedure. Some of these physicians apparently had more of a problem abandoning the idea of bleeding patients than others, going so far as to claim that industrialization or even changes in the earth's magnetic field had led to the recent ineffectiveness of bloodletting.

By the 1930s the need for *Hirudo* husbandry dropped off considerably—and it pretty much stayed that way until the mid-1970s. In New York City and Boston, pharmacists sold leeches through the 1920s, mostly to treat black eyes. But generally their use as a therapeutic tool became limited to a very few rare conditions and most leech breeders were forced to find new jobs.

"Then in the 1970s," Rudy said, "a few surgeons started using leeches after reconstructive surgery—especially in cases where they were doing reattachments. Some of these guys picked up the technique after serving in Vietnam and now they wanted a source for leeches back home."

Rudy explained how his business partner, Marie Bonazinga, had met up with Jacques des Barax, the president of Ricarimpex, at an international conference on medicinal leeches in the early 1980s. Not long after, Leeches USA was formed and surgeons in the United States and Canada would have easy access to an ancient but valuable surgical assistant—one that required little if any training and could be delivered by overnight express, courier, or even helicopter. Currently, Ricarimpex sells around seventy thousand leeches a year.*

Rudy recounted a number of cases—some of them rather high profile—in which leeches from his company played an important role in surgical reattachments. One such case involved a young female musician who was pushed off a New York City subway platform and had one of her hands severed by a passing train. The woman's hand was successfully reattached, and although she could no longer pursue a career as an instrumentalist she went on to become a successful occupational therapist.

Another was the case of John Wayne Bobbitt, who gained an unwanted degree of fame when his wife, Lorena, briefly took the penal code into her own hands. Driving off with a tip from her husband, Mrs. Bobbitt eventually tossed the evidence out of her car window. Remarkably, the misplaced member was recovered by a sharp-eyed policeman and Rudy's company was called upon to supply the leeches necessary for what was to become a combination reattachment/enlargement surgery.

In addition to their importance in reattaching structures like fingers and ears, a hundred or more leeches might be used over the course of several weeks in cases where the scalp has been

*Not everyone is enamored with the therapeutic uses of leeches, as was recently illustrated by the development of an artificial leech by researchers at the University of Wisconsin. Basically, the device is a glass vacuum chamber with separate tubes for suction and the introduction of the anticoagulant heparin. Inserted just under the skin, the disk-shaped tip rotates to inhibit blood clotting.

accidentally torn from a person's skull. Typically, though rarely, this occurs when the victim's hair is pulled into a piece of heavy machinery.

Leeches are also commonly employed in breast reconstruction, although in 1993, a leech went missing after such a surgery. Concerned physicians eventually determined that the wayward creature had entered the sutured wound. The wandering worm was later recovered from inside the patient's breast.

Leech therapy is also used in transfer operations, when skin or muscle is grafted from one place on the body to another. Two Slovenian surgeons, who reported their results in the *British Journal of Plastic Surgery* in 1960, undertook the first of these leech-assisted transfers. Some thirteen years later, the procedure would save my father's leg.

In July 1973 my father was involved in a terrible boating accident while we were on vacation in upstate New York. The propeller of our ski boat motor tore apart his knee, leaving a horrible wound and little for surgeons to use to reconstruct the joint.

To replace the missing flesh surrounding his knee, a multistep procedure started when a flap of tissue from my dad's abdomen was partially removed, then stitched into a tubelike structure. After a week or so, one end of the tube was detached from his body and reattached near his wrist (the other end of the tube was still attached about eight inches to the side of Dad's belly button). Once the circulation was reestablished, the belly end of the tube was removed and reattached to the area just below his elbow—so that it now resembled a handle spanning his forearm.

In this type of surgery—where a flap of skin is "walked" to a new destination—leeches are commonly used if circulation in the transferred skin begins to fail (i.e., if the color of the transferred skin changes from pink to purple).

"Check it out, Billy," I remember my dad saying. "They're turning me into a set of luggage!"

I could tell that my mother was absolutely mortified at this weird-looking handle of flesh—although she hid it well from my dad.

"I think it looks kind of neat," I told her later, as we left the hospital.

"You would," she said, shaking her head.

After my dad's "handle" stabilized, one end was detached from his abdomen and sutured onto his thigh, just above the damaged knee. An excruciatingly uncomfortable two weeks later (he had to remain with his left arm attached to his upper thigh by an eight-inch tube of flesh), the arm end of the handle was transferred to the area just below the knee.

My father never complained once during this whole painful and grueling process—one that ended a full three years after his accident with a surgery that basically fused his knee joint together. Within a month of returning home, Dad was chomping at the bit, and several months later he was back at work, driving an oil truck again.*

Besides their ability to draw off accumulating blood at surgical reattachment or transfer sites, researchers are currently exploring the possibility that leech "saliva" contains a virtual cornucopia of pharmacologically active compounds, including antihistamines and antibiotics.

"*Hirudo* and some of its relatives may offer alternative treatments for ailments ranging from osteoarthritis to the circulatory problems associated with diabetes," said Rudy Rosenberg.

*Less than three years after returning to work, my dad had a major stroke that crippled his body and brain for the last thirteen years of his life. William A. Schutt Sr. died in the spring of 1992 at the age of seventy-one. He had survived the poverty of the Great Depression, D-day, a nightmarish boating accident, and a horribly disabling stroke. He was the bravest man I have ever met.

The latter was a condition that Rosenberg had faced in the early 1980s after he received a call from his sister in San Diego.

"She said that one of my mother's legs was completely discolored and that doctors wanted to amputate it immediately. She was eighty-two years old at the time."

Rudy recounted how he instructed his sister to hold off on approving the surgery until he got there. Then he talked to his mom about an idea he had.

"I asked her if I could try to restore the circulation to her leg by using leeches. She agreed and off I went."

Rudy plopped twelve leeches into a jar filled with distilled water, sealed it tight, and headed to the airport, having booked the first flight to San Diego.

"When I got out there I could see that her leg was in bad shape. It was nearly black, and when I tried to apply the first leech, it refused to bite."

By now I was leaning forward, literally perched at the end of my chair.

"I abraded the skin a bit, and about fifteen minutes later the second leech finally took hold. Within ten minutes her toes started to turn pink."

"That's incredible," I said.

"Yes, that's what I thought too. I wound up treating her for three days."

"And . . . ?"

Rudy grinned, glancing up at the framed photo of a smiling woman seated on a chair. She could have passed for Albert Einstein's mom. "And she lived to be ninety-seven—with *both* of her legs."*

*Rudy actually finished the story by reiterating that leech therapy shouldn't be undertaken at home or without the supervision of a physician.

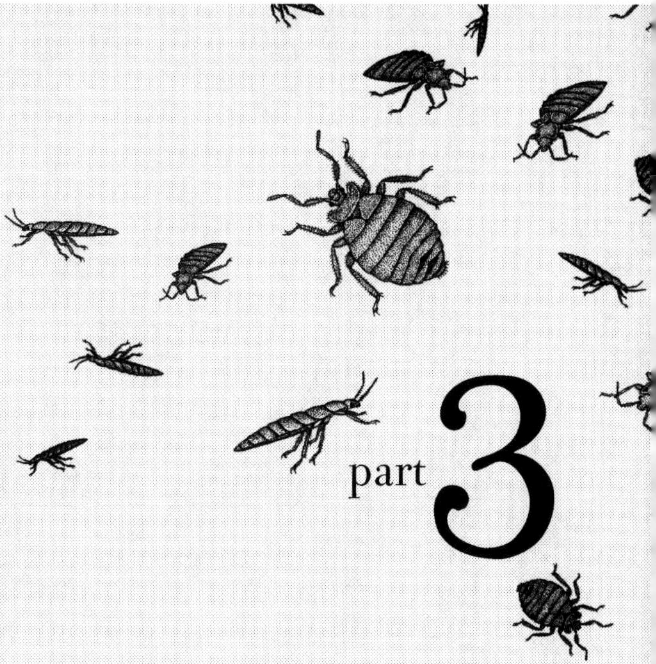

part 3

Bed Bug & Beyond

Kitanda usicho kilala hujui kunguni wake. (You cannot know the bugs of a bed that you have not lain on.)

—Swahili proverb

There may be bugs on some of you mugs
But there ain't no bugs on me.

—Wendell Woods Hall, "Ain't Gonna Rain No Mo"

With, ho! Such bugs and goblins in my life.

—*Hamlet,* act V, scene 2

7.

SLEEPING WITH THE ENEMY

Louis Sorkin reached across a table stacked high with cardboard specimen boxes and glass vials of every size. There was also enough Tupperware to throw a party for the entire Upper West Side of Manhattan. He shared the combination office/lab space with another entomologist, although collaboration would have been a challenge since the men worked on opposite sides of an island of insect-related paraphernalia that had risen from the fifth floor of the American Museum of Natural History to claim the middle two-thirds of the room.

"Did you want to have a look at my bed bug colony?" Lou

inquired, a few minutes after I'd arrived. He could have been asking if I wanted to see his kid's latest school photo.

"Definitely," I replied, leaning in from my seat as he snagged what looked to be a fist-sized canning jar before handing me the metal-covered container.

The first thing I noticed was that the bottom of the jar was padded with a generous wrapping of duct tape. Another band of tape prevented anyone from unscrewing the metal lid. In the center of the lid, a circle the size of a quarter had been removed and covered with a layer of fine mesh. I would later learn that this was a section of plankton net that had been secured to the ceiling of the colony by a generous schmear of silicon glue.

An air hole, I guessed. To my horror, I would later discover that I was only half right.

Inside the jar, there was a gray cardboard baffle, folded like an accordion. I tilted the container slightly and took a closer look—the cardboard was flecked with tiny black spots but there was no movement. *Nobody home.*

"I can't see anything." I said, even though, for a second there, I thought I might have seen *something*—a shift in the darkness between one of the cardboard folds.

"Cup the jar in your palms," Lou instructed, holding his hands as if in prayer.

I complied, although now it was impossible for me to see the interior of the container. About fifteen seconds later the entomologist nodded in my direction. "That ought to do it."

I shifted the canning jar to my left hand, bringing it closer to my face so that I could peer—

"Holy shit!" I screamed, and for a second, the jar shifted precariously in my hand. I secured my grip, then held the container at arm's length.

The entire inner surface of the jar was seething with movement—tiny flat ovals—some the size of apple pits, others more like sesame seeds, and all of them frantically pressing themselves against the glass. More and more of the creatures appeared. Within seconds there seemed to be *hundreds* of them pouring out of that single piece of folded cardboard.

I sensed Lou coming up behind me.

"Look closer," he said.

I squinted. There *was* something else. Amid the shifting pits and seeds were minuscule dots—barely visible and noticeable only because they were showing quite a bit more determination than your common household dust speck. In fact, if anything, their movements were even more frantic than the "giants" that clambered around them.

I found myself checking the silicon seal on the jar, having quickly come to the realization that a thin wall of glass and glue was the only thing keeping the bed bugs from the object of their frenzy—*me.*

"I'll let you feed them later, if you like," Lou said, almost as an aside.

"That would be great," I said, having absolutely no idea what I'd just agreed to do.

Several days earlier, I'd contacted Lou because I was interested in learning what was behind the recent and dramatic resurgence in bed bugs, the epicenter of which appeared to be New York City.* It seemed that every week the local papers were featuring stories about people attacked in their sleep by the tiny blood feeders, but the weird thing was that these attacks weren't taking place in rundown apartments or "no-tell motels." The rich and famous were being bled as they exercised in posh fitness centers and as they slept in ritzy Riverside Drive co-ops. And not only were they getting bitten up and grossed out (sometimes enduring

*According to New York City's Department of Housing Preservation and Development, there were no complaints of bed bugs in fiscal year 2003, 79 complaints in 2004, 928 in 2005, 4,638 in 2006, 6,889 in 2007, and 9,200 in 2008.

hundreds of bites) but these folks were also starting to squawk about it, as only New Yorkers can squawk. Guests at upscale hotels, both at home (the Helmsley Park Lane, overlooking Central Park) and abroad (the five-star Mandarin Oriental in Hyde Park, England) were filing huge lawsuits, not just because they'd suffered bed bug bites but also because of the cimicid souvenirs they'd brought home with them.* By January 2007 things seemed to have reached a fever pitch. There were splashy front-page stories in places like the *Village Voice* ("Bed Bugs & Beyond") and major articles in the *New Yorker* ("Night Visitors") and the *New York Times* ("Everything You Always Wanted to Know About Bed Bugs . . ."). Web sites (often offering conflicting information) and bed bug–related blogs sprang up on the Internet, some of them logging thousands of hits each month. Even politicians were getting into the act, scrambling to enact legislation to prevent the sale of secondhand mattresses. In typical fashion, the media gravitated toward a few experts who could combine bed bug–related expertise with a good sound bite (and occasionally, what my aunts would have considered to be some damned strange behavior). In that regard, American Museum of Natural History entomologist Lou Sorkin and classical music programmer–turned–exterminator/therapist Andy Linares had quickly become bed bug superstars.

What was causing the uproar? Why had bed bugs come back with such a vengeance, and where had they been for the past fifty

*In the Helmsley case, a Mexican businessman alleged that he'd been mauled by bed bugs during his stay at the hotel. The case was quickly settled out of court. The Mandarin Oriental case, however, did not go away quietly, possibly because the plaintiffs were a prominent New York City celebrity attorney and his wife. As of March 2007, their lawyers had filed a twenty-page five-count complaint alleging that the couple had suffered over a hundred bed bug bites in an assault that continued after they returned home and the bed bugs moved into their Manhattan apartment. The couple was seeking over four million dollars in damages—which works out to around forty thousand dollars a bite.

years or so? How were they spreading so rapidly and what could be done about them? And just what were bed bugs anyway? That seemed like a decent question to start off with.

Back in Lou's office I continued to stare at the creatures that were now trying desperately to eat me. "Are there different species of bed bugs in here?" I asked, having noted that the insects seemed to come in several handy sizes.

"No, they're all the same species, but you're looking at six different developmental stages."

I would soon learn that the smallest members of Lou's colony were the "first instars"—newly hatched bed bug nymphs, eager for their first blood meal, and nearly invisible until they had gorged themselves.

For members of the phylum Arthropoda (which includes insects, spiders, scorpions, crabs, lobsters, and shrimp), growth presents a different set of challenges than those encountered by vertebrates like mammals. Mainly, this is because the arthropods' hardened skeleton is located on the *outside* of its body. Additionally, rather than having adjoining bones articulate at specialized surfaces, their joints are actually composed of thin, highly flexible, sections of exoskeleton.

This jointed structure produces movement in its owner much like its vertebrate counterpart (with pairs of muscles working in opposition to each other)—except that in arthropods the muscles are found *within* the skeleton rather than outside of it.* Since exoskeletons don't grow once they've hardened, in order for the juvenile arthropod to attain a larger body size, the entire skeleton has

*This muscle is actually what you're eating when you chow down on the succulent meat inside a lobster, crab, or shrimp.

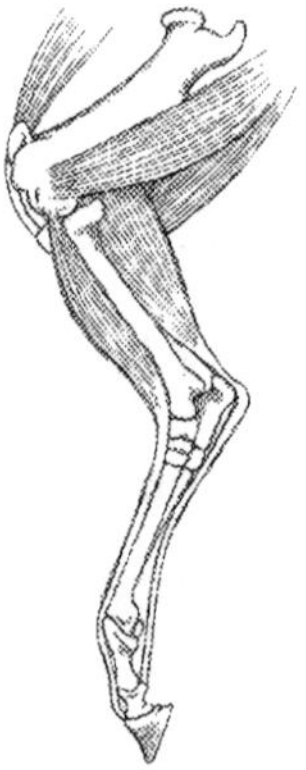

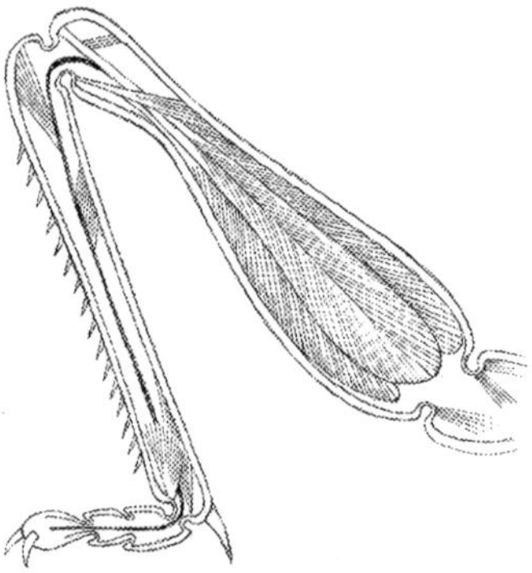

to be shed periodically. Ecdysis (Greek for "escape" or "slipping out of") reoccurs at the end of specific developmental stages called instars, eventually culminating at adulthood.* In some arthropods (like moths and flies), the early instars (caterpillars and maggots, respectively) don't resemble the adult stage at all. These strange-looking eating machines are referred to as larvae (or the larval stage). In other arthropods (like bed bugs and many other types of insects), the instars are called nymphs, and each successive nymphal stage more closely resembles the adult form.

The colony of *Cimex lectularius* I was holding in Lou Sorkin's office contained all five nymph instars plus the adult, or reproductive, stage. Each of these developmental stages was succes-

**Ecdysiast* was also the term coined by the early-twentieth-century reporter and critic H. L. Mencken (famous for his coverage of the Scopes "Monkey trial" of 1925). Mencken had been contacted by Georgia Southern ("a practitioner of the art of strip-teasing") who was concerned about negative connotations people had about "stripping." Ms. Southern asked Mencken to "coin a new and more palatable word to describe this art" and he did. "I sympathize with you in your affliction," Mencken wrote back to her. "It might be a good idea to relate strip-teasing in some way to the . . . zoological phenomenon of molting . . . which is ecdysis. This word produces . . . ecdysiast."

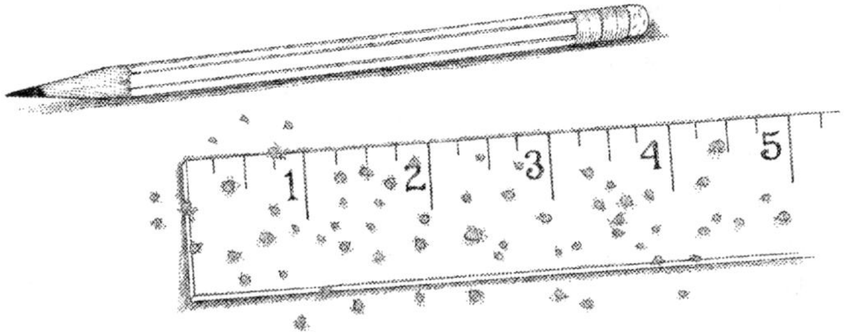

sively larger and dependent on obtaining a blood meal that would swell a body to the point that it would burst through a suddenly ill-fitting exoskeleton. Enzymatic secretions and an increase in blood pressure also helped to split the bed bugs' outer cuticle, which is composed of a tough waterproof polysaccharide called chitin. Littering the crevices where they hide (like outgrown clothes from Gymboree), I would come to learn that the sloughed-off skeletal casts were one of the telltale signs of a bed bug infestation.*

*Molted arthropod casts and the feces dropped by their former owners are the real allergens that plague those who are hypersensitive to "dust." Although I'm sure there are a few allergists who have chosen to spare their patients some of the following details, the discarded arthropod outerwear and microscopic excreta come from dust mites, tiny creatures more closely related to spiders than to insects. Fortunately for us, dust mites are not blood feeders (although, as we'll see later, there are hundreds of mite species that are). Instead, they feed on the approximately twelve grams of skin flakes that humans shed each day. This epidermal debris would accumulate into gigantic snowdrifts were it not for the hungry mites.

Arthropods that have recently undergone ecdysis avoid predation by hiding out until their soft-shelled armor solidifies.*

One final note about this soft-shelled phase: it's been hypothesized that structural collapse of the arthropod body is a potential hazard for those squishy-limbed individuals who have recently emerged from a round of ecdysis. This rationale was then used to explain why the largest aquatic arthropods (like lobsters and king crabs) are *far* heavier than their terrestrial counterparts (crabs, insects, spiders, and centipedes).† Since water is more viscous than air (i.e., it's thicker), objects in water are supported to a greater degree than they are on land (where gravity is much more of an issue). According to this rationale, heavyweight arthropods going through their soft-shelled phase can support themselves only if they live in water. Another way to look at it is that size in terrestrial arthropods is limited (i.e., constrained) by two physical parameters: gravity and viscosity. Beyond an interesting explanation for

*When I was a child, soft-shelled crabs were expensive and available for only a few weeks each summer. What I didn't know was that they were actually just ordinary blue crabs *(Callinectes sapidus)* that either didn't get the good hiding places after a molt or had been tempted to leave them by the irresistible allure of rotting chicken entrails. Nowadays though, farm raised soft-shelled crabs can be purchased year-round since their molts are regulated by human-administered hormones.

†The largest terrestrial arthropod, by far, is the coconut crab *(Birgus latro)*, which can weigh up to 8.8 pounds (four kilograms). *Birgus* is a hermit crab relative but without the snail shell. It copes with the "soft-shell vs. gravity" problem by molting in the safety of its burrow, where it remains for up to thirty days. Depending on which entomologist you ask, the largest insects are either New Zealand grasshoppers, called wetas, which can weigh as much as 2.5 ounces (seventy grams), or giant beetles belonging to several genera like *Goliathus, Titanus,* and *Megasoma.* The world's largest spider (class Arachnida) is the giant bird-eating tarantula *(Theraphosa blondi)*, which can weigh up to 4.25 ounces (120 grams) and whose legs can span twelve inches (30.5 centimeters). Its venomous fangs can reach one inch (2.54 centimeters) in length. Some centipedes (class Chilopoda), like the Amazonian giant centipede *(Scolopendra gigantea)*, can reach more than twelve inches in length. These predators feed on pretty much anything that moves—including rodents and bats, which are subdued with a venomous bite before being devoured. The heaviest lobster on record is forty-four pounds, six ounces (about 20 kilograms), which is five times heavier than the coconut crab and roughly three hundred times heavier than the largest insect.

arthropod size differences, to me the *real* take home message here is our tendency to believe that in evolution *anything* is possible. In reality, though, constraints like those placed on arthropod body size by gravity and viscosity serve to illustrate that in nature some forms (like car-sized bed bugs) just aren't possible.*

As I rotated the canning jar in my hand, I couldn't help noticing that Lou's bed bugs were in a frenzy.

"They're attracted to your body heat and the carbon dioxide in your breath," Lou said.

Within the crowded glass confines, the bed bugs had responded to stimulatory cues in much the same way as the leeches did when hunting Bogie's character in *The African Queen*. In this case, however, the thermoreceptors and chemoreceptors responsible for prey detection were stimulated by increases in temperature and carbon dioxide concentration (rather than touch, from the incoming waves of disturbed water, or vision, from changes in light intensity). At a basic level, though, the wiring of the bed bug and leech nervous systems, and their function, is quite similar. A stimulus is detected that prompts signals (afferent nerve impulses) to be sent from sensory receptors to the body's data processing center (the brain). After rapidly sorting through incoming information (like direction of prey and distance), a response is generated that takes the form of outgoing (efferent) nerve impulses. These are sent to the muscles of locomotion. Activation of these muscles and their subsequent contraction leads to coordinated movement of the leech or bed bug's body (either swimming or running toward its

*Those of you looking for a more detailed explanation on the concept of evolutionary constraints should read Stephen J. Gould and Richard Lewontin's terrific essay, "The Spandrels of San Marcos and the Panglossian Paradigm: A Critique of the Adaptionist Programme."

respective prey). In both instances, if the initial stimulus had been interpreted by the brain as DANGER, rather than FOOD, the outgoing response would have resulted in defensive behavior–like fleeing.

This is certainly a simplification, but on one level the only difference between the nervous systems of leeches, bed bugs, and humans is that we have *many* more neurons, packed into specialized regions of the brain (like our wrinkly cerebral hemispheres). This complex and intricately interconnected wiring allows us to do things that the relatively simplified nervous systems of the leech or bed bug cannot achieve–such as deciding whether to respond to a stimulus in the first place or choosing to vary that response. In the previously mentioned blood feeders, fewer neurons lead to limited or even one stereotypical response to each stimulus encountered. For example, bed bugs are thought to release an aggregation pheromone, a chemical that initiates clustering behavior in members of the same species (conspecifics).* In this case, think of a trio of bed bugs that have just hopped off someone's luggage. After scurrying across the floor for a few seconds, one of them encounters a wall and follows it, eventually finding a crack in the molding large enough to slip through. Stimulated by physical contact with the walls of this dark, safe haven (hereafter referred to as a harborage), the bed bug releases a pheromone whose message is interpreted by the other two bed bugs as something very much like SAFEDARK.

Initially released in response to a stimulus, the pheromone itself becomes a stimulus, triggering a highly specific response. Soon enough, the harborage has three bed bugs in it. Killing time while they await a meal, the bugs behave predictably, making

*Pheromones are sprayed or otherwise released into the environment by a variety of creatures, including insects and many mammals. Different pheromones communicate information on territorial boundaries, availability of females for breeding, and location of trails (to food or back to the nest).

more little bed bugs and producing copious piles of bloody feces.*

The point is that once pheromones are sensed, there's no choice and little behavioral variation in the response. These chemical messages are also a major key to the seemingly bewildering degree of organization exhibited by social insects like ants, termites, and bees.

Although Lou's bed bug colony was a far cry from a beehive, the single-mindedness of their quest for blood was chilling to watch.

"Imagine having this many bed bugs living in your apartment—living behind your headboard, in your mattress, hiding behind your switch plates."

"And all of them just waiting for the lights to go out," I chimed in. Admittedly, I was starting to buy into Sorkin's ghoulish gig.

"Exactly," Lou said. I noticed that there was nothing that could be interpreted as disgust in his voice, and I wondered how many people this soft-spoken bug-meister had sent home to a night of the creepy-crawlies after they'd checked out his colony.

He went on. "And the more cluttered your home is, the better."

I shot a quick glance around Lou's office. "So what you're saying is that I should not drop this bottle."

"No . . . that would be a bad thing," the bug expert replied.

I passed the jar back to the researcher, but instead of placing it back on his desk, he did something peculiar. He brought the lid of the jar up to his nose and inhaled (rather deeply, I thought).

"Some people say they smell like fresh raspberries or cilantro." He held the bottle toward me.

*Bed bug copulation is actually quite dangerous for the female since it entails a violent exercise known as traumatic insemination. During this process, the male pierces the female's abdominal wall with his intromittent organ and injects his sperm into the wound. This practice, in which the female's external reproductive structures are not involved at all, may have evolved as a way to circumvent female mating resistance. Not surprisingly, traumatic insemination has some negative effects on female bed bugs, increasing the risk of infection and reducing life span and reproductive output.

I took a small sniff.

"Hmmmm," I said, not smelling much of anything.

"I always thought they smelled like citronella," Lou continued. "Nowhere near as strong as those yellow citronella ants, but there's a definite similarity."

I leaned over and took a somewhat larger hit, checking first to make certain we hadn't inhaled a hole in the mesh. They *did* smell like citronella.

The scientist motioned to my note pad. "This is important," he said. "There are Web sites and articles out there reporting that bed bugs don't have a smell. That's untrue—*especially* when they get riled up."

I nodded, as I took some notes. Strangely, rather than the smell of bed bugs or their lack of smell, I'd been struck by the idea that someone other than myself had actually sniffed those citronella ants. Each summer when I was a kid, the quarter-inch-long insects would swarm within the sidewalk cracks in front of my house. They had a distinctive odor, and as soon as I caught wind of it, I'd break out my action figures and start fishing for ants with a piece of straw. The enraged ants would emerge, a dozen or so at a time, clamped by their powerful jaws to the NEST THREAT that appeared like clockwork each year to poke at the entrance to their colony. And just like Lou's bed bugs, the angrier the citronella ants became, the stronger the scent they produced.

The entomologist's voice jerked me back into the present. "In all likelihood, bed bugs release many different pheromones."

Besides chemical messages like the aggression pheromone produced by the citronella ants I'd hassled as a kid, other substances released by the bed bugs functioned to make them less palatable to predators.*

*Various insects and other arthropods prey on bed bugs. These include several species of ants, including Pharaoh ants *(Mnomorium pharaonis)*, a bug called the "masked bed bug hunter" *(Reduvius personatus)*, spiders *(Thanatos flavidus)*, centipedes *(Scutigera*

"Humans can only discern one of these pheromones," Lou continued, placing his colony down on the table. "Dogs, on the other hand, have a much more sensitive sense of smell, and some are actually being trained to detect bed bug infestations."*

At a lecture sponsored by the New York Entomological Society, several nights later, I learned that researchers were working to identify the bed bugs' aggregational pheromone—the chemical signal that would lead to the formation of loose groups by the bed bugs as they gathered in their nooks and crannies between meals. By isolating the specific chemical that caused bed bugs to aggregate, scientists hoped to learn more about this behavior—information that could be used to develop more effective eradication methods.

There were around seventy-five people present in the audience that night and they seemed to be a mix of pest-control types (there to pick up half a New York State Department of Environmental Conservation credit for their attendance) and city dwellers, either interested in bed bugs or traumatized by them to various degrees of twitchiness.

The lecture was titled "Good Night, Sleep Tight, Don't Let the *Cimex lectularius* Bite," and the first speaker was Dr. Jody Gangloff-Kaufmann, a former doctoral student in entomology at Cornell and currently working for New York State's Integrated Pest Management program. Together with her cospeaker, Gil Bloom (a faculty member at the City University of New York), she had been tracking the current outbreak of bed bugs in New York City since early 2001.

According to Dr. Gangloff-Kaufmann, bed bugs originally lived

---

forceps), and pseudoscorpions *(Chelifer cancroids).* Finally, although an 1855 paper reported that cockroaches fed on bed bugs, this claim has not been supported and the two insects may, in fact, "live happily together in the same house."

*The owners of Advanced K9 Detectives (out of Milford, Connecticut) claim that their certified Bed Bug Dogs can sniff out infestations within minutes.

in caves and fed on bats. Once humans (and other mammals) began inhabiting these caves, the opportunistic parasites began to feed on them as well. At a certain point, some bed bugs became associated rather exclusively with humans.*

The first literary reference to bed bugs can be found in Aristophanes' play *The Clouds* (423 BCE). A century later, in *Historia Animalium,* Aristotle assured readers that "bugs are generated from the moisture of living animals, as it dries outside their bodies."

Monograph of Cimicidae, by Robert Usinger, is the closest thing to a bed bug bible. In his book, Usinger describes how bed bugs were not only present in Greece by 400 BCE but that the Greek physician Dioscorides was advising patients to *eat them.* For example, a recipe that called for mixing seven wall lice with meat and beans was used as a treatment for malaria. By "holding the beans," one could counteract the venom of certain snakes.† For those who preferred to take their ectoparasites with a chaser, Dioscorides also prescribed downing the bugs with wine or vinegar as means of expelling horse leeches (presumably from a patient's throat). Additionally, difficult or painful urination (a

*Parasites switching hosts is a common occurrence in nature—and it's one of the ways that new species emerge. Recently, scientist David Reed and his co-workers compared the DNA in species of sucking lice that prey on either humans or gorillas. They hypothesized that gorillas actually spread these blood-feeding parasites (arthropods belonging to the order Anoplura) to ancient human ancestors around 3.3 million years ago. Since that time, the lice evolved alongside their new hosts (i.e., underwent coevolution), eventually becoming different enough to be considered a separate species from the gorilla lice. For example, as humans lost most of their body hair, the lice became adapted to live in thatches of human pubic hair, where, unlike gorilla lice, they are transmitted mainly through sexual contact. The researchers believe that the lice were originally spread to humans via one of three routes: sexual contact between gorillas and early humans, ancient humans killing and handling gorillas (parasites often spread from one host to a predator of that host), or gorillas and humans sharing communal areas.

†Bed bugs 'n' beans was apparently a popular medicinal dish (showing up in Pope John XXI's *Thesaurus Pauperum*). But here, rather than mixing the two ingredients together, those suffering from fever were instructed to place the bugs into a hollow bean before swallowing it.

condition called dysuria) was treated by mashing up some of the insects and inserting them into the stricken orifice. Even *sniffing* them (the bed bugs, that is) could revitalize a woman who had fallen into a swoon from "strangulation of the vulva."

Medicinal uses for bed bugs, most of them cribbed from the ancient Greeks, were described in 77 CE by the Roman Gaius Plinius Secundus (better known in modern times as Pliny the Elder).*

Quintus Serenus was another Roman savant and the author of an early medical text. Here, the author clearly demonstrates how his savant status had been attained in fields other than poetry writing:

Shame not to drink thee Wall-lice mixt with wine,
And Garlik bruised together at noon-day.
Moreover a bruis'd Wall-louse with an egge, repine
Not for to take, 'tis loathsome, yet full good I say.

Serenus does fare slightly better in this next description of an alternative Roman thirst quencher (actually cracking several Top 100 lists of "poems about consumption of bloodsucking insects").†

Some men prescribe seven Wall-lice for to drink,
Mingled with water, and one cup they think
Is better than with drowsy death to sink

With no reports on any actual medical benefits derived from eating, drinking, sniffing, or inserting bed bugs, it appears that

*This was two years before Pliny (a scholar, historian, and naturalist) choked to death on volcanic gas and dust at Stabiae, shortly after the cataclysmic eruption of Mount Vesuvius.

†According to Usinger, both of the Quintus Serenus verses were quoted by T. Mouffet in 1634 and translated by E. Topsel in 1658.

their medicinal use was yet another instance of treatments that were nearly as bad as the maladies they were meant to alleviate (see chapter 4).

A Treatise of Buggs, written by John Southall, was the first book devoted completely to bed bugs. Published in 1730, it offers readers a tantalizing peek at early pest control, as well as some insights into race relations.

During a visit to the West Indies in 1726, the author was puzzled after encountering "an uncommon negro" with hair, breast, and beard "as white as snow." The old gentleman was also puzzled by *his* encounter with the author, noticing that his "Face and Eyes were much swelled with Bugg-Bites," and he wondered why "white men should let them bite," rather than doing "something to kill them, as he did." Presumably having no good answer for that one, Southall accepted "a Calibash full of Liquor" and directions to apply the stuff around his bedroom. The results would have sent a chill down Lou Sorkin's spine:

> The instant I applied it, vast numbers did, (as he told me they would) come out of their Holes, and die before my face.
>
> (John Southall, *A Treatise of Buggs,* 8)

After waking up bite-free, the author showed his gratitude by immediately hatching a plot to separate the freed slave (now referred to as "my Negro") from his secret recipe. Southall broke out the good stuff, enticing the Jamaican with "one piece of beef, some biscuits and a bottle of beer," after noting how "all Negroes being greedy of Flesh, when they can come at it." As the day progressed, the brew flowed freely until "all the bottles we emptied of beer were fill'd with liquor." Southall, however, remained sober enough to make notes about ingredients, quantities, and procedures, and after returning to England he marketed the pesticide along with his services as an early pest-control specialist. Unlike his long-forgotten Jamaican "business partner," Southall did not divulge the secret ingredients of his elixir—which he christened Nonpareil.*

Southall also endeavored to determine just how bed bugs came to England, and in doing so he shows off his modest side, informing the reader how he overcame "difficulties, which might have discouraged a less enterprising Genius." The great man consulted "as many learned, curious, and ancient men" as he could find, affirming that before the Great Fire of London in 1666, bed bugs "were never noted to have been seen."

> They were then so few, as to be little taken notice of; yet as they were only seen in Firr-Timber, 'twas conjectur'd they were then first brought to England in them; of which most of the new Houses were partly built, instead of the good Oak, destroy'd in the old.†
>
> (John Southall, *A Treatise of Buggs*, 17)

Southall's interviews supported the claims made in several early European dictionaries and encyclopedias that bed bugs did not exist in London prior to the Great Fire but were subsequently carried to England in timber imported from the American colonies. Years later, documents would reveal that the bloodsucking pests had actually been recorded in England since 1583 (nearly one hundred years before the famous London blaze).

Miffed Americans returned fire in the eighteenth century by nicknaming the tiny bloodsuckers "red coats" and insisting that their own bed bug problems had arrived from Europe with the early colonists. In this regard, the Yanks were apparently correct

*Hopefully the name referred to the French translation ("peerless") and has nothing to do with those chocolate drops covered with little white pellets of sugar.

†Firr is an Old English take on the word *fir* and refers to evergreen conifers of the genus *Abies*. These trees are generally considered unsuitable for use as timber.

since entomologists now believe that *Cimex lectularius* spread from an origin somewhere in the eastern Mediterranean region, across much of the world, via human colonization and overseas trade.

Finally, Southall set out to describe bed bugs for his readers—with decidedly mixed results:

> A Bugg's Body is shaped and shelled, and the Shell as transparent and finely striped as the most beautiful amphibious Turtle; has six legs most exactly shaped, jointed and bristled as the Legs of a Crab. Its Neck and Head much resembles a Toad's. On its Head are three Horns piequed and bristled; and at the end of their Nose they have a Sting sharper and much smaller than a Bee's. The Use of their Horns is in Fight to assail their Enemies, or defend themselves. With the Sting they penetrate and wound our Skins, and then (tho' the Wound is so small as to be almost imperceptible) they thence by Suction extract their most delicious Food, our Blood.
>
> (John Southall, *A Treatise of Buggs,* 19)

Currently, scientists recognize around seventy-five species in the family Cimicidae, but only three of them regularly feed on the blood of humans: *Leptocimex bouti,* which also preys on bats in western Africa and South America; *Cimex hemipterus* (sometimes known as the tropical bed bug), which feeds on poultry and bats in the New and Old World tropics (including Florida); and *Cimex lectularius,* the common bed bug, which preys on humans, bats, poultry, and other domesticated animals just about anywhere in the world.

Reflecting their worldwide distribution, Robert Usinger listed over sixty native names for bed bugs. Besides "red coats" and "heavy dragoons" (after the scarlet-coated British cavalry), additional English nicknames included "mahogany-flat," "B. Flat," and "scarlet ramblers." "Norfolk Howard" was a goof on the aristocratic family name of the Dukes of Norfolk, and in the first half of the twentieth century, the blood-filled hordes were known as "the Red Army." Bed bugs have also been referred to as "chinch bugs," probably because *chinche* is the Spanish word for bed bug. Unfortunately, this has led to some confusion since the name chinch bug has also been appropriated by the lygaeids, a related family of soil-dwelling insects notorious for the damage they inflict upon grasses and grains.

During my visit with entomologist Lou Sorkin, I asked him about the old perception that bed bugs were found only in association with hobos, seedy motels, and filthy conditions.

He shook his head. "That's been the mind-set for quite some time, but in the old days, only people who had money could heat their homes, so that's where you'd find the infestations. And once central heating took off, so did the bed bugs."

I glanced over at the colony. The jar was sitting on Lou's worktable, and now that the creatures within it weren't being warmed, breathed on, or sniffed, most of the miniature horde had retreated back into the shadows of their cardboard harborage.

The bug man went on to explain how increased temperatures not only attracted bed bugs and amped up their activity, but it also sped up their life cycle. "Higher temperatures lead to faster maturation to the adult, reproductive stage."

I would later learn from Dr. Gangloff-Kaufmann's presentation that a combination of high temperature (85°F) and high humidity could condense the bed bugs' entire life cycle into a period of three to four weeks. She explained that initially this might sound like a good thing, because the pests died more quickly, but they could also crank out a new generation in less time—leading to an overall increase in population size.

"So can cooling down an infested home get rid of bed bugs?"

"Not really," Lou said, shaking his head. "Lower temperatures can slow down their maturation process but it also increases their life span."

Like other insects, when bed bugs are chilled their metabolic rates decrease.

"Nymphs can go for months and months without feeding and adults can live without a blood meal for a year or longer."

Well, here was a clue I hadn't read about on Web sites or in the rash of recent newspaper and magazine articles on bed bugs. Rather than becoming a plus in the war against them, the creatures' adaptive response to low temperatures presented a significant problem: bed bugs could survive for months with no food, in empty (and presumably unheated) apartments. Harkening back to the enormous amount of misinformation on bats, the ability of bed bugs to survive prolonged periods without their human hosts apparently led to a fairly common belief that they could

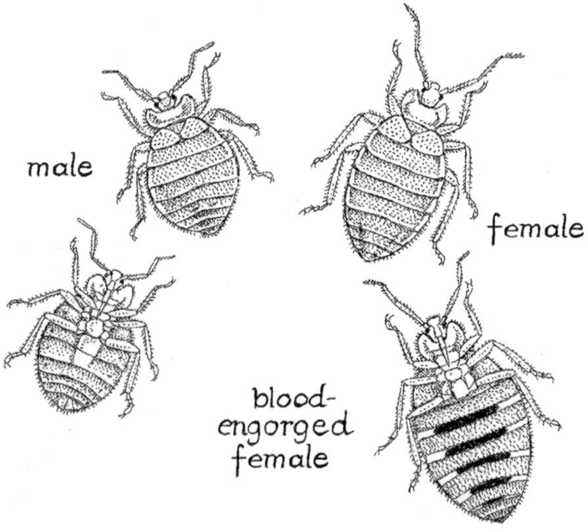

feed on juices extracted from wood and paper, or even digest wallpaper glue. "Paste they love much," declared self-proclaimed genius John Southall in his *Treatise on Buggs*.

I guess it might have been almost comforting to think that these tiny home invaders were munching glue or savoring newspaper ink. Comforting that is, compared to a pair of grim realizations that not only did bed bugs maintain a strict diet of blood, but unlike exotic vampires like bats and leeches, these hardened city dwellers were living (and feeding) right here among us.

Okay, so we already know that bed bugs are arthropods, like crabs and spiders, but let's get a bit more specific, starting with the question, what are "bugs"?

Bed bugs and their fellow cimicids belong to a large suborder of insects known as Heteroptera. These, in turn, belong to an even more inclusive grouping, the order Hemiptera. Although some hemipterans do feed on blood, many don't. Aphids, for example, the enemy of farmers and gardeners everywhere, feed on plant sap and cause *serious* damage in the process.

But no matter what they feed on, to be a card-carrying hemipteran, you need to have a needle-sharp, dual-channeled proboscis. After piercing the skin (or rind) of whatever it is they happen to feed on, hemipterans inject saliva through one channel of their proboscis. Compounds within the saliva begin the process of digestion, and almost immediately, the bugs start snorking up partially digested food through the other channel.*

*There are significant parallels between blood feeding and sap feeding in bugs. They both employ hypodermically sharp mouth parts to tap into nutritious fluids, pumped under relatively high pressure, through special vessels (blood and phloem vessels, respectively). Additionally, both techniques commonly transmit pathogens to the food source—harming them.

In addition to bed bugs, the Heteroptera contain a parade of insects with nasty-sounding names like stink bugs, squash bugs, and water scorpions. Assassin bugs (Reduviidae) are another notorious family of hemipterans. Unlike their bed bug cousins, though, some assassin bugs can transmit serious ailments to the humans they feed upon. In fact, these insect vampires (also known as cone-nosed bugs or kissing bugs) deposit feces containing the parasitic flagellate *Trypanosoma cruzi* onto their victim's skin. Itching the bug bite rubs the infected excrement into the wound, allowing the parasite to enter the bloodstream. From there it can invade organs like muscles. In severe cases, the ailment is known as Chagas' disease, in which the parasites can cause serious damage to nerves of the gastrointestinal tract and the electrical conduction system of the heart. Charles Darwin was, in fact, bitten by assassin bugs in South America, and it's been suggested that Chagas' disease may have been responsible for the lifelong health problems he experienced upon his return to England.

Although *bug* can refer to diseases like influenza, as well as just about any insect or small arthropod (e.g., spiders and ticks), only heteropterans are considered "true bugs" by entomologists. This is because they all share some rather specific anatomical characteristics. For example, many insects have two pairs of wings (forewings and hind wings). In most heteropterans, the forewings are hard at the base and membranous toward the tip (hence their name, from the Greek for "different wings"). One take-home message is that "all bugs are insects but not all insects are bugs."

Strangely enough, although bed bugs share anatomical, developmental, and behavioral similarities with other heteropterans, they don't have functional wings. The posterior wings of cimicids are absent and the anterior pair is vestigial. Vestigial organs are nonfunctional remnants of structures that *were* functional in the ancestors of that organism. For example, blind cavefishes (belonging to the families Amblyopsidae and Characidae) have tiny, functionless eyes. By all indications, these sightless swimmers evolved from ancestral species that *could* see. Presumably, some of these fish migrated into new environments (caves), where they eventually lost their visual senses in much the same manner as other troglodytes like blind salamanders and some cave crickets. The sightless eyes that remain result from portions of the fishes' genetic blueprint (its DNA) that have remained unchanged from the ancestral versions. As a result, these old sections of DNA are still cranking out remnants of the old anatomical features—even though they don't function anymore.* In bed bugs, one pair of stubby functionless wings (hemielytra) is all that remains of what were probably two working pairs of wings in bed bug ancestors. Presumably, the loss of wings occurred as these ancient bugs evolved a lifestyle in which birds and bats spread them from place to place, thus rendering their own wings unnecessary.

Speaking of bugs, the English word *bug* apparently derives from the Welsh *bwg*. In its original form, *bug* (or *bugg*) referred to a ghost or hobgoblin, which is how it appeared in the *Coverdale Bible* (1535) and in several works by William Shakespeare (including *Hamlet*).† In the 1622 play *The Virgin Martyr*, *bug* appears to have been used for the first time to describe an insect infestation ("We have bugs, Sir!"). Until then, *wall lice* and *Cimices* had been used to describe *Cimex lectularius*.

Of course, since the seventeenth century, the word *bug* has

*Other vestigial organs of note include hind limb claws (or spurs) in pythons and boa constrictors and teeth in the developmental stages of toothless baleen whales (like the blue whale) and some anteaters. In humans, there's the coccyx (formerly caudal or tail vertebrae in our distant ancestors) and the vermiform appendix (the remains of a larger, endosymbiont-laden, intestinal out-pocket called the cecum). Additionally, wisdom teeth are an indication that natural selection had an easier time making our jaws shorter than it did decreasing the number of teeth in those jaws. And stress-induced goose bumps apparently functioned to raise the hairs of our ancestors, making them look larger in the presence of predators, potential enemies, or competitors.

†The term *bug bear (Henry IV)* was an early term for "scarecrow" or "bogeyman."

developed a number of additional meanings. The phrases "putting a bug in someone's ear" or "having a bug up one's butt" may very well have originated with the medicinal use of leeches (but I'm only speculating here). Used as a verb, *bug* can describe certain unwelcome attentions or the covert placement of surveillance equipment. It can also refer to the equipment itself or a computer or technical glitch. Drinking "bug juice" means chugging an inferior grade of alcoholic beverage (except in ancient Greece, where the term was taken literally and seems to have been the equivalent of huffing down a bean-and-bug-flavored Alka-Seltzer). In any event, bug juice could leave the drinker "bug eyed" or as "crazy as a bed bug" and in no way capable of driving certain formerly ubiquitous German autos. Finally (and *thankfully*), the phrase "snug as a bug in a rug" first appeared in an eighteenth-century farce, *The Stratford Jubilee,* whereas "Sleep tight, etc., etc." has two possible origins. According to columnist Cecil Adams (the Straight Dope), "sleeping tight" may refer to a time when many mattresses were made of interwoven strands of rope attached to a rectangular wooden frame. To sleep well, these mattress ropes had to be pulled tight and reknotted (lest they sag like a hammock under the weight of the sleeper). The alternative explanation is that in this case *tight* refers to the word's archaic use as an adverb meaning "soundly," "properly," or "well."

Why do bed bugs find living with humans so comfortable? And just as importantly, what are *we* doing that makes it so easy for them to thrive and spread from place to place?

Let's begin with some housing and transportation issues.

Many ectoparasites (like the ticks, mites, and chiggers, which will be discussed elsewhere in this book) use an array of specialized appendages to cling or otherwise attach themselves to their

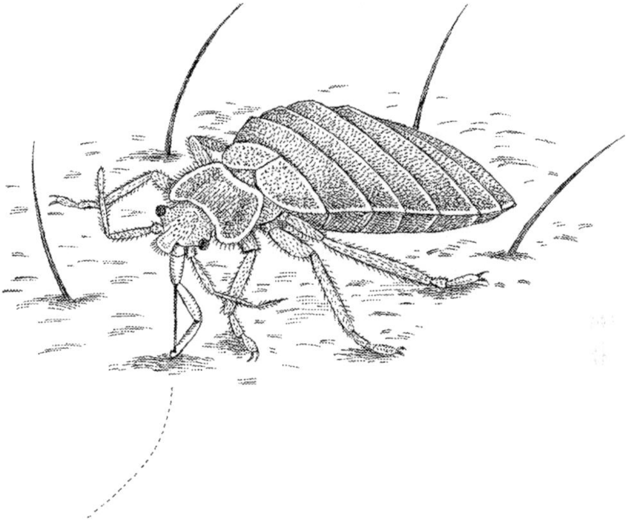

hosts—sometimes for extended periods. Bed bugs and their relatives, however, spend a major portion of their lives hiding in close proximity to their hosts but not living on them.* Generally, bed bugs react negatively to light and actively seek out rough, dry surfaces that are at least partially darkened. They emerge from these harborages late at night (usually around 3 or 4 a.m.), climbing aboard their prey for short periods of time to feed.

The common bed bug, *Cimex lectularius,* generally feeds for five to ten minutes, often making three or four bites, roughly in a row (sometimes referred to as "breakfast, lunch, and dinner") before returning to its harborage. After a feeding bout, bed bugs are ready to eat again within a week.

By the time a victim notices the bites, the bed bugs are long gone, leaving behind a cluster of itchy, red blotches, bumps, or

*To some researchers, this makes them blood-feeding predators and not parasites.

welts that can occur on any exposed skin surface (e.g., face, neck, shoulders, arms, hands, etc.). Reaction to the bites can vary since it depends on the victim's immune response to proteins in the bed bugs' saliva, but generally, the more bites, the greater the level of inflammation. Unfortunately, bed bug bites are often misdiagnosed by physicians as mosquito or flea bites or, more commonly, as scabies, an itchy skin condition caused by the microscopic mite *Sarcoptes scabei.*

Since birds are parasitized by a wide variety of ectoparasites (including many species of bed bugs and their relatives), bird nests are often infested with tiny ectoparasites (like chiggers, lice, and ticks). Nests provide the miniature vampires with the perfect microhabitat for activities like breeding, hiding out, and waiting for the home delivery of their next meal.*

Similar to cimicids that prey on birds, those that feed on bat blood spend the majority of their lives killing time in places where their prey hang out—in this case, literally. Bat roosts generally vary by species but they're commonly located in caves, mines, attics, abandoned buildings, and tree hollows. There, the bat bugs hide in cracks and crevices—quietly digesting their meals and presumably catching up on the latest bits of misinformation about their hosts.

With this behavior in mind, it's no surprise that bed bugs have made an easy transition from bird nests and bat caves to the vast

*Remember this little factoid the next time you come across a bird's nest and are tempted to bring it home to show your kids. And, on a related topic, any newly dead animal is likely to be crawling with ectoparasites—and many of them will gladly jump off their deceased hosts and on to you (at least temporarily), should the opportunity arise. You should also think about this one the next time your cat brings home a recent kill.

heated structures created by humans and packed with potential hiding places of every conceivable shape and size.

"Clutter is the bed bugs' best friend," said bed bug expert Gil Bloom, during his presentation. And so, in many human homes bed bugs have found paradise.*

Similarly, the actual dispersal of bed bug colonies (aka spread of the bed bug infestation) has developed a significant human element.

According to Bloom, there are two ways that bed bugs can be introduced into a home: actively and passively. In active introduction, bed bugs migrate from one place to another under their own power. Since bed bugs don't have functional wings, active dispersal of the colony depends on walking (or running) to a new home.

Migration from one room to another is easy enough to visualize but what about between apartments or floors? As Bloom tells it, bed bugs can easily relocate within a building via pipes, phone wires, or cables.

Humans can often pick up an assist in these active introductions and here's how. Let's say your neighbor upstairs has figured out that his apartment is infested with bed bugs. He decides to ditch his bug-filled bedding curbside and proceeds to wrestle his mattress out into the hallway. Some of the bed bugs might fall off as he tips the mattress on its side, while others are jostled off as the bugged bed is dragged down the hall or thumps its way down the stairs (think of that scene in *King Kong* with the sailors

*According to pest-control experts, developing hobbies like dusting and vacuuming are solid preventive measures, while storing stuff under your bed is *bad juju*. They also recommend sealing all cracks and crevices in your home with silicon or caulking compound. (Potential harborage sites for bed bugs include cracks in hardwood flooring, bed frames, moldings, as well as the spaces between walls or floors and semipermanent structures like bookshelves.) You should also paste down or replace peeling wallpaper or contact paper, fill in nail holes, and seal all holes in floors and walls where bed bugs might enter from outside. And before you sit down, check your toilet paper tubes and curtain rods.

clinging for their lives to a giant log as Kong tries to shake them off). Rather than falling to their deaths, though, the displaced vampires hit the ground running, heading for the first crevice they can find. In all likelihood, this means scooting under doors and into new apartments. Once they get themselves settled (think about how those aggregational pheromones work), females will start pumping out five or so eggs a day (several hundred in a lifetime) and a new colony can form almost as fast as you can say, "Honey, check out this red spot on my arm."

The role of your neighbor in this scenario leads us to the second method by which bed bugs can infest a residence. Passive introduction pretty much covers any transport method that *doesn't* employ the bed bugs' own locomotor abilities. In cimicids that do not feed on humans, this generally occurs when bugs are delivered via airmail to new locals by unsuspecting bats or birds. In *Cimex lectularius,* passive introduction relies primarily on humans—their products and their wonderful efficiency at moving from one place to another. As we'll see, this ability to exploit our habits as well as the things we use on a daily basis has become one of the major factors in the current spread of bed bug populations.

Let's say your neighbor has succeeded in humping that mattress down five flights of stairs (and potentially spreading the bed bugs to five new floors) before dumping the thing curbside. If a college student or someone in the market for cheap bedding picks it up, the bed bug infestation will start spreading as soon as the new owner lugs the mattress into his or her apartment. It may even spread to *her* neighbor's apartments as the mattress gets dragged up a new set of stairs and down yet another hallway.

But what if that scenario never takes place? What if people are smart enough *not* to pick up someone else's old bed? Perhaps the curbside mattress has been labeled by its former owner as being infested with bed bugs. In that case nobody in his or her right mind would touch it, right? Unfortunately, this just isn't the case—

and not by a long shot. All too often, discarded mattresses and box springs are quickly snagged by companies that specialize in collecting and "reconditioning" these old mattresses before offering them for resale. According to several sources (who wish to remain unnamed), these secondhand mattress companies send around trucks manned by sharp-eyed crews. Their job is to pick up any box springs or mattresses they encounter—*even those that are clearly marked by their former owners as harboring bed bugs.* Supposedly, these bedding items are then "sanitized" before being "rebuilt." But according to one New York City exterminator, unless "sanitizing" means baking at 150°F for forty-five minutes, or treatment in a fumigation chamber, the bed bugs present in the mattress and their eggs are not killed.

Fortunately, in some states like New York, there are laws to regulate the sale of reconditioned mattresses. Unfortunately, the lack of enforcement guidelines means that nobody is out there checking to see what these secondhand vendors are doing. Rather than being effectively sanitized, it's far more likely that used mattresses get a quick disinfectant spritz and a new cover—right over the old one.* So, in addition to bed bugs (which can be hidden throughout the mattress), the old mattress might be contaminated with urine, saliva, and just about anything else your imagination can dredge up. After a quick turn-around, the refurbished bedding is sold to a generally clueless public (many of whom probably think they're getting a new mattress, perhaps with a bit of old hardware within).† In instances where the recycled and resold bedding is already infested with bed bugs, the results are predictable.

*Pest-control experts do suggest that you cover your mattress with an airtight, hypoallergenic cover, which not only prevents bed bugs from getting out (if you did have them) but also prevents new ones from getting in.

†According to the U.S. Federal Trade Commission (FTC), the easiest way to determine whether a mattress is new or refurbished is to look for an attached label. New mattresses should have a white label stating that it contains "all new materials." Depending on the state, used mattresses *may* have a tag warning consumers that the

Here's another serious, bedding-related problem that helps explain the current bed bug resurgence. Let's say you've done your homework and found a bedding store with a stellar reputation. You pick out a mattress and box spring and set up a delivery date. So far so good—but here's where things start to unravel. Typically, in addition to delivering your new mattress, the company will haul away your old one—and even though *your* old backbreaker might be completely bug-free, there's a real possibility that some of the mattresses these guys pick up during their workday are going to be infested with bed bugs. These contaminated mattresses are going to be placed onto the same truck as your new mattress—perhaps they'll even be leaning against each other. And what about the interior of the delivery truck (dry, dark, and full of cracks and crevices) or even the deliverymen? Are you going to check their cuffs for tiny specks of bed bug feces? Some of these guys might have something to say about that (and the ones who don't could be even more of a problem).*

So the next time you purchase a new bed, you just might be getting more than only a new mattress and box spring delivered to your home (*especially* if you happen to be one of the day's last deliveries).

Alarmed by the spread of mattress-related bed bug transmission, New York City councilwoman Gail Brewer introduced new legislation in September 2006 that would ban the sale of all re-

mattress contains used materials. In New York, for example, sellers of used bedding must attach "a 15 square inch yellow label . . . which contains the phrase 'Used Material' in prominent print." With this in mind, purchasing a mattress with no tag should be avoided and the FTC recommends that you not let the heavy plastic wrapping dissuade you from locating the tag when your mattress is delivered. The FTC further suggests that refusing delivery of untagged mattresses is a wise move, as is getting the salesperson to write "new" on your receipt at the time of purchase.

*According to integrated pest-control specialist, Dr. Jody Gangloff-Kaufmann, as of 2007, one major bedding company covers the mattresses it removes from homes in plastic before placing them in their trucks.

conditioned mattresses. Introductory Bill Number 57 (or Intro. 57 for short) also required that new and used mattresses be transported separately. Unfortunately, while this law would be a move in the right direction, it wouldn't come close to stopping the bed bug problem.

"They sort of forgot to include box springs," said Andy Linares, the owner of Bug Off Pest Control in northern Manhattan.

"What about couches, futons, and nightstands?" I asked.

"Whoops," Andy said, smiling.

And *that* was a problem. How many thrift-conscious city dwellers have snagged a piece of furniture or other household items from a secondhand store or off the curb? Bed bug harborages aren't limited to mattresses and box springs, and it is unlikely that a change in the city's administrative code would have much of an effect on bed bug populations, if any. The bottom line is that people should forget about bringing *anything* into their homes that they've picked up curbside. They should also be very

careful about what they purchase from secondhand stores, flea markets, or obtain from furniture rental outfits.

According to Lou Sorkin, "They can hide just about anywhere. Clock radios, TV remotes, telephones, picture frames, lamps, headboards—basically any type of furniture." Books and wall hangings are also popular aggregation destinations as are the spaces behind the switch plates covering electrical outlets.

The bed bugs' ability to adapt to our methods of transportation is another reason for the recent resurgence in these creatures. Remember that in the days before civilization, the spread of bed bugs was limited by the range of their bird or bat hosts (and the bug's ability to hold on and not look down). Before the twentieth century the majority of people lived and died without traveling much and the resulting spread of bed bugs was gradual. But once humans began traveling, bed bugs followed them. Nowadays bed bugs are rapidly dispersed, sometimes over enormous distances, by the potentially millions of people who routinely travel in cars, buses, trains (subways), and planes.

Dr. Tamson Yeh, an entomologist at the Cornell Cooperative Extension, had her own hypothesis as to how bed bugs were getting around big cities like New York.

"Taxis," she told me during a visit to her office in Riverhead, Long Island. "People put their bags or suitcases down next to the curb and bed bugs can climb right on—or climb right off."

"Jeez," I chimed in, "and just think about how many people are traveling all over the world and then tossing their suitcases into the trunks of cabs when they get home."

"It's the perfect environment," Tammy said. "Dark, dry, plenty of places to hide . . ."

"And how often do cabbies sanitize their trunks?"

"Exactly."

Once limited to movement from cave to cave or nest to nest, as bed bugs became associated with humans, it was no stretch for them to migrate from room to room or apartment to apartment. Now, however, *Cimex* infestations are spreading across cities, between states, and even to different countries. So right up there among the multiple reasons for the twenty-first-century resurgence in bed bugs, cheap, fast, long distance transportation is near the top of the list.

According to pest-control expert Andy Linares, "Sometimes outbreaks can be traced to overseas travel since it's pretty easy to pick up bed bugs from cruise ships, resorts, hotel rooms, or hostels."

"My cousin's kid just got back from a hosteling trip through Australia," I said.

Andy shook his head. "The media is full of reports of people bringing back bed bugs after spending big bucks at top-notch hotels or spectacular resorts. And hostels? That's scary."

The bug man went on to explain how basically any place that has a high resident turnover rate (e.g., shelters, dorms, hostels, hotels, and apartment buildings) pretty much fits the bill as a potential source for passive introduction.

"So where are the trendy bed bugs heading nowadays?" I asked.

"Eastern Europe is a hotbed of bed bug activity," he said. "And England has a huge problem."

Apparently, a suitcase or backpack, opened or even set down in an infested room can serve as a sort of bed bug version of the Trojan horse. And on a related note, by placing your clothes into a hotel dresser drawer in an infested room, you can also easily pick up some unwelcome traveling companions.

To minimize the risk, Andy recommended that travelers examine their rooms *before* bringing in their luggage and other belongings. Although this might sound a bit extreme, he stressed the following preventive measures as the *least* you should do: Start your search at the corner of the bed nearest to the clock alarm. Carefully lift up the sheet and the mattress cover and examine the mattress, especially around buttons or along the raised seam. Using a flashlight if necessary, look for fecal stains (tiny, dark-colored raised bumps) or for the bed bugs themselves (flattened apple seeds with legs). Then lift the corner of the mattress and look at that section of the box spring. Use your flashlight again to examine the space between the headboard and the wall. If you're still suspicious, look under pillows and inside pillowcases. Bed bugs can live in the clock alarm, the nightstand, or even the bedside lamp. If you find any bed bugs or even any fecal stains (which appear as pinhead-sized raised dots, usually dark brown in color), leave immediately and insist on another room. Of course, you should repeat your inspection in the new room and be prepared to "bail" on the hotel, if need be. Finally, while traveling, keep luggage elevated off the floor and check it carefully for unwanted hitchhikers. Hard plastic suitcases are more resistant to bed bugs than fabric suitcases with their multiple nooks, creases, and folds. In any event, once home, you should thoroughly vacuum your luggage and store it in sealed, black plastic bags—but not in your bedroom.

This illustrates another distressing point. Even if you don't bring bed bugs into your home, it's quite possible that someone else might. Plumbers, electricians, visiting nurses, and house cleaners can become involved in passive introduction. And when guests arrive at our homes, how many of us throw their coats and handbags onto our beds?

Bed bugs are turning up in hospitals, doctor's offices, health clubs, and movie theaters. They can be transmitted to your clothing if you sit on infested furniture or happen to brush up against someone wearing bed bug–laden clothes.

Along similar lines, Gil Bloom suggested in his talk that people should be a bit cautious when a friend asks to stay at your apartment for a few days "for unspecified reasons." I guess the implication here is that your jokey reply, "You're not having your place treated for bed bugs, are you?" should *not* elicit a coughing fit, nervous laughter, or sudden and profuse sweating.

"Bed bugs can break up relationships and friendships," Andy Linares told me. "And even people who don't have them can get screwed up."

"How's that?" I wondered.

Andy explained that many people mistake specks of dirt or lint for bed bugs.

"I tell them to stay calm and not to freak out. Use a magnifying glass. Lint doesn't have six legs, and it doesn't crawl around by itself."

I nodded. I'd have to remember that one.

"Sometimes they think they've been bitten by bed bugs, but it's not a bite at all. In those cases, I'll ask them if there's a construction site nearby," Andy said.

"Why's that?"

"A lot of people have allergic reactions to concrete dust. If that shit gets on your skin it will definitely make you itch."

Some individuals, however, experience a much more serious problem than concrete dust. The unfortunate people suffering from delusional parasitosis (sometimes referred to as Ekbom's syndrome) believe that they're being plagued by parasites crawling over their bodies and sometimes under their skin. These imaginary pests appear to vary with the individual—snakes, insects, and other vermin are often reported. Sufferers will also claim that these creatures are infesting their homes, clothes, and belongings. In some cases, delusional parasitosis results from drug use ("cocaine bugs" or "meth bugs") or extreme alcohol withdrawal, but recent paranoia over bed bug infestations in places like New York City has apparently amped up some city dwellers to a new level of neurosis.

During lunch one afternoon, Dr. Jody Gangloff-Kaufmann recounted an incident that occurred while she was working for the Nassau County Cooperative Extension. "A guy came in convinced that there were bugs crawling all over his body. To prove it to us, he brought in his bedsheets *and* his underwear."

I made a face that brought our waiter running over (probably thinking I'd found a bone in my sushi).

"Yeah, that's how we reacted," Jody said, as I waved off the waiter.

"What did you find?" I asked.

"We didn't find anything," she said. "But apparently that didn't satisfy the guy. He wound up spraying down his entire body with a garden pesticide—his ten-year-old kid too."

"The symptoms are always the same," Andy Linares told me. "People report an itching sensation but no bites. They see stuff crawling around. The next day it's flying. They come in with blobs of fabric and pieces of belly-button lint."

"Andy, these people who imagine that they've got bed bugs—do you ever treat their apartments, just to get them off your back?"

"Absolutely not!" the bug man exploded. "If you do that you'll *never* get rid of them."

He explained how a background in diplomacy and international affairs (he holds a master's degree from Fordham University) has actually helped him cope with frantic victims of bed bug infestation—both real and imagined.

"People can deal with roaches and rats, but bed bugs are another story. They're ninja insects—cryptic and insidious—and people feel powerless and violated by them."

A big part of his job, he said, was to act as a sort of mental health counselor. "It's almost therapeutic when these people put themselves into my hands. They're stigmatized by old ideas about bed bugs, poverty, filth, and such. So they come in undercover. 'Don't let anyone know what you're doing,' they tell me."

At that point, my conversation with Andy was interrupted by a phone call. It was a woman who had just figured out that her daughter had brought back bed bugs from her college dormitory. Now the tiny monsters had infested her home and the woman was frantic. He'd call her back in a few minutes, he told her.

Andy shook his head. "She's going to freak when I tell her what she needs to do before we can treat her house."

"Why's that?" I asked.

"Because she'll have a lot of homework to do before we get there. Furniture, bedding, magazines—you've got to be ruthless about throwing stuff out," Andy told me. "Everything that has a crack, crease, or crevice has to be chucked out, steam-cleaned, or vacuumed."

"Everything?" I asked.

Andy nodded. "People with huge libraries or old LP collections are pretty much screwed."

After the prep work, "infestibles"* are either packed into a sealed container to be heated or pumped full of a fumigant like Vikane (sulfuryl fluoride) for forty-eight hours.

"Bed bug infestations were quite common before World War II," Andy Linares explained, "but in the 1940s and 1950s, the use of DDT pretty much put a stop to them."

Paul Müller, the Swedish chemist who figured out that DDT was an effective contact pesticide against arthropods like mosquitoes, ticks, and moths, was awarded the Nobel Prize in physiology. Mueller discovered that DDT worked by causing neurons to fire spontaneously—not a helpful thing if you're trying to fly, bite, or crawl.

*Basically, this refers to all of your property except pets, plants, and food.

By the mid-1950s, however, bed bug resistance to DDT had became so widespread that two alternate pesticides, malathion and lindane, became the tools of choice for controlling them. Unfortunately, during the next decade, studies began to show that the lethal effects of these pesticides weren't just confined to the insects.*

"Pyrethroids are our tool nowadays," said Dr. Jody Gangloff-Kaufmann, during her lecture.

Pyrethroids are man-made compounds, similar in chemical structure and insect killing properties to the natural pesticide pyrethrum, produced by chrysanthemum flowers. According to a fact sheet issued by the Illinois Department of Health, "When used properly, pyrethroids have been found to pose very little risk to human health and the environment." Unlike the previously mentioned (and banned) insecticides, pyrethroids apparently break down within a day or two of application. This means that should you ingest some, it won't stick around in your body to cause problems like birth defects or cancer.

Not bad, I thought at the time. *Sounds easy enough.*

But several days later, I learned from Andy Linares that there was nothing simple about the treatment of bed bugs, and that pest-control specialists were in fact using a number of additional substances in their war against cimicids.

The Bug Off Web site, for example, suggested that exterminators might inject "a variety of flushing agents (565-XLO, CB123 Extra), aerosols (D-Force), liquid residuals (Permacide Concentrate, P-1 Quarts, P-1 Gallons), powders (Drione), sanitizers (Ster-

*As it turned out, all three of these compounds were extremely toxic to humans and most other forms of life. As vividly portrayed in Rachael Carson's landmark book *Silent Spring,* pesticides like DDT caused long-term environmental and health effects. Eventually, the alarm raised by *Silent Spring* over DDT would launch the modern environmental movement, and by the mid-1970s malathion, lindane, and other compounds known as organophosphates and carbamates were banned from use as pesticides in the United States and other countries.

ifab Pints, Sterifab Gallons, Sterifab 5-Gallon), and growth regulators (Gentrol Aerosol, Gentrol Vials, Gentrol Pints), in all possible cracks and voids as part of a comprehensive treatment."

"Bed bugs are becoming resistant to pyrethroids," said American Museum of Natural History entomologist Lou Sorkin.

Andy Linares agreed. "Using a variety of chemical compounds minimizes the resistance factor."

Dr. Gangloff-Kaufmann also addressed the question of bed bug resurgence with another treatment-related explanation. "In the old days," she said, "exterminators used to regularly spray baseboards and molding with pesticides to control roaches."

She went on to explain that, in all likelihood, these sprays not only killed the roaches but also had a positive effect on bed bug control.

"But roach control is different nowadays," she said. "In many instances, they use poison baits instead of spraying. And since bed bugs only feed on blood, roach or ant baits are ineffective."*

According to Andy Linares, "Those older sprays offered additional protection because they would actually vaporize and redeposit themselves onto areas adjacent to where they'd been sprayed initially."

One of the things *everyone* seemed to agree on is that "bug bombs" are a *terrible* idea for dealing with bed bugs.

"They're one of the worst things you can do," said Dr. Gangloff-Kaufmann.

"You might kill some individuals, but you'll send others into voids and sheltered areas—and possibly into your neighbor's apartment," warned Gil Bloom. "And with bug bombs, most of the bugs won't get a lethal dose anyway."

*On a related note, "sticky traps" used to trap insects like cockroaches don't work against bed bugs, which are more likely to travel and hide *under* the trap than run across its gooey surface.

"It'll just irritate them and get them to move around a bit," Lou Sorkin told me as we sat in his museum office.

I asked the entomologist if there were any other reasons why bed bug treatments themselves had become part of the problem.

"Well, for one, in New York State, you can't do a preventative treatment on a building, which means that, by law, you can't treat a place for bed bugs until there's a reported infestation."

"How come?"

Lou hesitated, and I could tell that he was a bit uncomfortable answering this particular question. But after some wheedling on my part, I was able to determine that the reasons for banning preemptive strikes against bed bugs (as well as certain pesticides) were more politically based than scientific.

"Hey, politicians have to keep their constituents happy," Lou quipped.

I decided to leave it at that, suddenly remembering Lou's earlier offer to have me feed his bed bug colony.

I nodded toward the canning jar. "So how *do* you feed those guys, anyway?"

"Simple," he said. "You just invert the jar and hold it against your arm for five minutes or so."

As I watched, Lou rolled up his shirtsleeve. I couldn't help noticing that there were several circular patches of red skin on his forearm and that each was the exact size and shape of the mesh-covered hole in the jar lid.

"That's really . . . neat," I sputtered.

"The welts don't really itch that much," he said. "And, anyway, I'm used to it."

He must have seen me staring at his arm. "Sometimes I guess I let them feed a little too long," he said with a shrug.

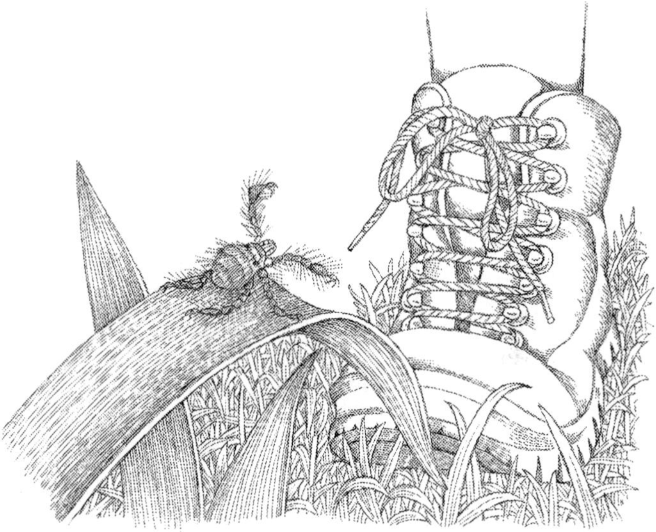

Ticks: the foulest and nastiest creatures that be.

—Pliny the Elder

That which is not good for the beehive
cannot be good for the bees.

—Marcus Aurelius *(Meditations)*

8.

OF MITES AND MEN

The first time I visited Trinidad in August 1991, I brought along one pair of long, cotton field pants and five pairs of shorts. I figured, Hey, it's hot there every day. Who needs long pants?

On subsequent visits to the country, or any other tropical locale I happened to be working in, I'd bring *one* pair of shorts, reserved for walking around town (if there was a town) and *five* pairs of long pants.

The reason for the drastic wardrobe revision can be summed up in one word: *chiggers.* As I found out, the easiest way to experience these tiny parasites firsthand is to wear short pants (and sandals) while walking through any grassy or wooded area.

Unfortunately, that's exactly how I encountered them as I hiked the forest trails behind the PAX Guest House where I was staying.

It hadn't seemed like a bad idea—at least initially. It was probably about 90°F with humidity to match and I'd snuck away from my lab work, thinking that a walk under a canopy of green might cool things down a bit.

The hike itself was uneventful and in that regard it was entirely different from my evening treks through the forest. Missing were the night sounds—layer upon layer of chirps, buzzes, and clicks, all of them set against the incessant drone of mosquitoes. With each nighttime walk in the rain forest came a growing awareness that the trees themselves were alive and covered with life, and this awareness brought with it a mild claustrophobia that was impossible to describe and never entirely went away.

But now, in the blast furnace of midafternoon, the forest was silent. Nothing moved.

Soon enough, I headed back to PAX, convinced that any creatures worth seeing were also smart enough not to be out and about in such wretched conditions.

After getting buzzed in past the front door, I was confronted by Gerard, the guesthouse manager. Gerard runs on the low side of the "Aunt Rose Height Scale," but he is a force of nature. Wildly funny and incredibly bright, he and his wonderful Dutch wife Oda run the mountaintop landmark as if they were born for the job. Gerard's only real personality flaw is that he hates bats (which is rather unfortunate for him since there are probably thirty species patrolling the rain forest right outside his back door).

"And where have you been, young man?" Gerard inquired, his voice rising to a pitch that threatened to crack my sunglasses. (Gerard refers to all males younger than the age of eighty-eight as "young man," and on a recent return trip to PAX, my wife Janet and grad student Maria were disappointed to learn that Gerard had been calling them *both* "sugar plum.")

"Just taking a walk," I said, trying not to drip on the exquisitely polished wooden floor.

Gerard shot me a look as if I'd said I'd been out munching road kill. "Whatever," he said, throwing me a good-natured but dismissive wave before hurrying past—presumably in search of more intelligent company.

The next day I awoke to find that not all of the forest creatures had been inactive during my ill-advised walk. Looking down, I could see a band of itchy, red dots circumnavigating my waist like Magellan's diaper rash.

After taking a hot shower, I slathered on some calamine lotion, but strangely, the itching just got worse so that by that evening I had run through the entire bottle, exhibiting all the self-control of an addict on hot pink liquid crack.

Embarrassed, I decided to keep my new red belt a secret.

"I hear you picked up a few chigger bites," my graduate adviser and mentor John Hermanson mentioned over breakfast the following day.

Great, I thought. It must have been my question about whether calamine lotion came in a five-gallon drum.

"Yeah, no problem," I said, with as much nonchalance as I could muster. "I'm feeling a lot better."

"That's good," he said, returning his attention to a plateful of eggs. "Because they'll probably arrest you if you keep rubbing against the table like that."

Over the next few days, the rash actually got worse, as did my desire to scratch the angry welts that had developed. I tried to get creative, imagining how I'd deal with this if I'd been stranded on an island somewhere. I found that by hooking my thumbs into my belt buckle and imitating some old twist meister, I could scratch all of my chigger bites simultaneously. Almost as important, I learned that I could avoid bites altogether by wearing lightweight long pants with the cuffs tucked into my boots. And as for protecting my upper body, I found that an ultralight, long-sleeved shirt added the finishing touch to a relatively chigger-resistant field outfit. The key bit of knowledge, though, was *not* to go wandering around the forest or scrub during the hottest, most humid part of the day.

Doing some investigative work, I determined that the itching and redness I'd been suffering through was a type of dermatitis called thrombidiosus and that I'd been lucky the problem hadn't stuck around for ten days instead of five or six. I was also lucky that the chiggers in Trinidad weren't transmitting some sort of nasty, trip-ending bacterial infection. Had I been trekking through the brush in Southeast Asia or the South Pacific, I would have run the risk of contracting "scrub typhus" from chiggers such as *Leptothrombidium akamushi*. The Japanese first described this tiny arthropod over two thousand years ago (*akamushi* is Japanese for something akin to "dangerous bug") but although *Leptothrombidium akamushi* isn't a bug, or even an insect for that matter, its bite can transmit a potentially lethal bacterium to humans.

Initially, the bacterium *Orientia tsutsugamushi* is inoculated into the skin of a host through a chigger bite or contact with the chigger's feces. It then spreads through the host's bloodstream where it invades the endothelium, the flattened layer of cells that compose the inner lining of vertebrate blood vessels. *Orientia* (a close relative of *Rickettsia rickettsii*, the bacterium responsible for Rocky Mountain spotted fever) gains access to the endothelial cells' interior in a rather dodgy way, and in doing so, it pulls off a microscopic version of the famous Trojan horse trick. Phagocytosed by the host cells, *Orientia* is packaged inside enzyme-filled death baggies called phagosomes. But instead of waiting around to be lysed (or "sliced up," for those of you who skipped the blood chapter), the bacterium thwarts the host's defenses by escaping its membrane-bound prison to take up residence in the cells' gel-like cytoplasm. There, *Orientia* multiplies rapidly by a form of asexual

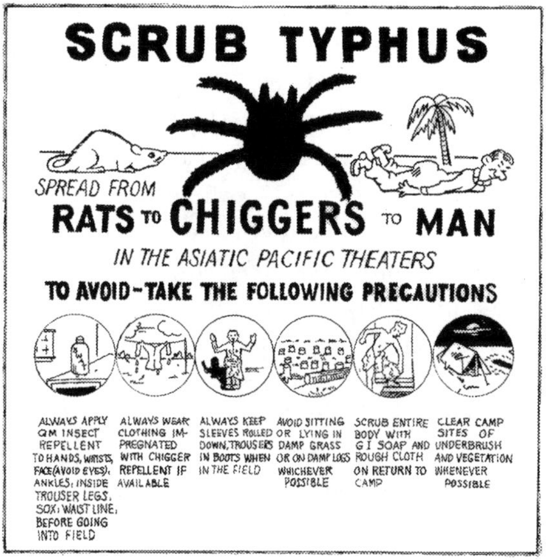

reproduction called binary fission.* Soon the bacteria-packed host cells burst, sending millions of new pathogens to infect new endothelial cells downstream.

During World War II, Allied forces in Southeast Asia and the South Pacific were ravaged by a number of chigger- and tick-borne diseases. Of these, scrub typhus became the most common as well as the most deadly. Sometimes described as mite-borne rickettsia (the rod-shaped bacterium was formerly known as *Rickettsia tsutsugamushi*), scrub typhus is characterized by high fever, muscle pain, painfully swollen lymph nodes, delirium, and a severe rash. If left untreated or treated too late, scrub typhus can cause encephalitis, circulatory failure, and even death.

In a pattern that's become familiar to those studying arthropod-transmitted diseases, rodents and not humans were the preferred

*Essentially, you start with one cell, its genetic material divides, its cytoplasm divides, and you end up with two identical daughter cells.

host of the pathogen-carrying chigger. In the Pacific theater of World War II, rat, vole, and field mouse populations exploded owing to the massive influx of troops, the garbage they produced, and the filthy conditions they endured. As a result, human chigger bites became more and more frequent, until scrub typhus reached epidemic levels. With no specific treatment available during the early 1940s, the disease killed more soldiers in the Burma-India-China theater of operations than any other infectious disease.* So serious was the scrub typhus problem that in some of the scientific literature from the 1950s, the chigger responsible for transmitting the disease was described in terms usually reserved for human enemies:

> All along the Papuan coast and adjacent islands Akamushi continued its ambush tactics against American troops. Invariably combat troops in most of the engagements of the island-hopping road to Tokyo sought the ever-present cover afforded by the tall kunai grass where Akamushi waited in hiding.
>
> (Emory C. Cushing, *History of Entomology in World War II,* 80–81)

Not surprisingly, the death toll from scrub typhus generated increased concern about the parasite causing the disease, and this concern led to an intense period of research that ushered in the modern science of acarology (i.e., the study of chiggers, mites, and ticks). Eventually, antibiotics like tetracycline, doxycycline, and chloramphenicol became effective treatments for scrub typhus and other acarid-transmitted diseases.

As with other bacterial pathogens, antibiotic resistance is be-

*According to the *Scrub Typhus Information Paper,* published in 2005 by the U.S. Army Medical Research and Material Command, there were 5,441 cases of scrub typhus, resulting in 283 deaths among U.S. Army personnel in the Asia-Pacific theater during World War II.

coming more of a problem. In parts of northern Thailand, for example, doxycycline-resistant and chloramphenicol-resistant strains of *Orientia tsutsugamushi* have evolved, and as a result, 15 percent of patients who contract scrub typhus die from the disease. This type of antibacterial resistance can be attributed to several factors: the tremendous rate at which bacteria multiply, the high rate of bacterial mutation, and the misuse of antibiotics.

In that regard, one of the most common ways that antibiotics are misused occurs when a prescribed antibiotic regimen is abandoned, usually once a patient feels better. Few people, in fact, seem to realize the danger posed when they decide to stop their antibiotic treatment before it's complete—and here's why. Think of a hypothetical population of 1,000 microbes inside a person who has been instructed to take an antibiotic for seven days. Discounting bacterial reproduction for a moment (since this is a model), let's say the antibiotic kills 900 microbes by day five and 990 by day six. If the patient were then to stop taking the antibiotic after day six, which microbes would be left alive? The ten survivors who were the most resistant to the antibiotic in the first place. *Now* factor in bacterial reproduction—and as these surviving microbes begin to multiply, each new generation will have the same antibiotic resistance exhibited by the ten original survivors.[*]

Now that we've seen some of the nasty effects of chiggers it's time to figure out just what they are. The short answer is that they're the parasitic juvenile stages of some mite species.[†]

*Using this same rationale, it's easy to see why bootleg antibiotics or antibiotics of unknown quality are a very bad thing.

†In addition to differences in diet, chiggers are smaller than adult mites and have six legs rather than eight.

And what about ticks?

Like chiggers, their story has become an extremely important one because of the pathogens they transmit through their bites. Starting in the mid-1970s, there has been growing concern in the United States and elsewhere over tick-transmitted diseases—especially Lyme disease and Rocky Mountain spotted fever.

Are ticks (like bed bugs) a pest on the rise, and if so, why? Are tick-transmitted pathogens also becoming immune to our treatments, or is there another answer? And what about Lyme disease? Why are the symptoms so variable (ranging from minor annoyance to catastrophic and life altering)? Some people believe that there's a chronic version of the disease—one that the experts refuse to talk about. And while we're in conspiracy mode, whatever happened to Lymerix, the Lyme disease vaccine?

All right, before we deal with the grassy knoll (and the chiggers lying in wait there), let's cover some basics. First of all, chiggers,

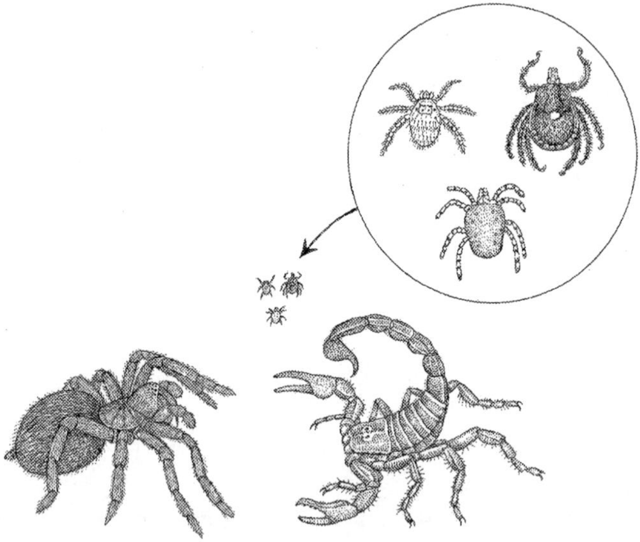

mites, and ticks are *not* insects, but like insects, they belong to the enormous invertebrate phylum Arthropoda. In fact, they're members of the only arthropod group that rivals the insects in diversity—Arachnida (a subphylum that also includes the spiders and scorpions).

While chiggers aren't *exactly* blood feeders, they merit interest because their blood-sucking cousins, ticks, share much of their biology as well as their penchant for strange behavior. Additionally, hundreds of mite species are vampires as adults. It's just that many of them feed on nonvertebrate blood—like hemolymph, the blood found in arthropods like insects. Some of these mite/insect interactions have also been in the news lately, especially as they relate to agriculture and the honey bee industry.

Evolutionarily, mites and ticks are *extremely* close. In all likelihood, tick ancestors (prototicks) were actually mites that somehow evolved to become obligate blood feeders (similar to vampire bat ancestors, which were non–blood feeders).

Why did some mites retain their larval blood-feeding lifestyles into adulthood while other body characteristics (like sexual organs) developed normally? The answer is that maintaining larval (or juvenile) characteristics as an otherwise mature adult is yet another example of how new species can evolve. The basic premise, proposed by the evolutionary embryologist Gavin de Beer (1930) and reinvigorated by Stephen Jay Gould in his book *Ontogeny and Phylogeny* (1977), is that "evolution occurs when ontogeny* is altered in one of two ways: when new characters are introduced at any stage of development with varying effects on the subsequent stages, or when characters already present undergo changes in developmental timing."

In the first scenario, Gould's "introduced characters" result

*Ontogeny is the development of the individual from fertilized egg to adult.

from genetic alterations like mutations—changes in an individual's genetic blueprint that occur during DNA replication.* It is at this time that errors (variations to some) in the copying mechanism result in mutated strands of DNA. In this classic explanation of how evolution operates, in some instances this mutant DNA results in new characteristics for the individual.†

We've already seen the hypothetical results of such mutations (horses and vampire bats), but let's look at the second of these examples a bit closer. Let's say that a protovampire bat had a genetic mutation that resulted in a change in tooth structure. If this mutation happened to produce sharper teeth (giving the protovampire a better chance of biting animals without being detected and thus increasing its chances of surviving and reproducing), then this novel characteristic would be considered an "adaptation." Subsequent generations of protovampires, which would include the progeny of the bat with the mutation, would exhibit greater incidences of sharper teeth since those protovampires without this trait would be less likely to survive and reproduce in their local environments. In time, protovampire populations might accumulate more and more new characteristics (like salivary anticoagulants and amped-up excretory systems), adaptations similarly "selected" by the existing environmental conditions. Eventually, these bats would become different enough from their ancestors to be considered a new species—in this case, true vampire bats. Alternatively (and this is the part that many people overlook), the mutation that altered tooth sharpness could have just as easily been one that produced duller teeth, and this "mal-

*DNA replication (i.e., DNA synthesis) is the process by which the double-stranded DNA molecule is copied before cell division (resulting in two identical double-stranded DNA molecules and thus ensuring that each of the two resulting cells contains a full copy of the original genetic blueprint).

†Alternately, as we'll see, mutations that affect the *sequence* of developmental events can also provide variation that serves as the raw material for evolutionary change.

adaptive" character would have lessened the chances of that individual surviving to reproductive age, where it could pass that mutation (and the trait) on to the next generation.

Considering this last possibility makes it easier to understand how evolution is, in some ways, a genetic crapshoot. It also allows us to recognize the problem with such assumptions as "The environment changed and ancient vampire bats *needed* sharper teeth to make painless bites—so they evolved sharper teeth." This reasoning, which seems to make sense, betrays a common misconception that people have about the mechanism of evolution.

For Jean-Baptiste Lamarck (1744–1829), the first naturalist to propose a mechanism for evolutionary change, the concept of need-based evolution (and other related ideas) got the Frenchman buried in a rented grave and his life's work all but forgotten. In reality, Lamarck was a scientific heavyweight, and his résumé included prescient insights into botany, taxonomy, and organic evolution. He may have been, in fact, the first scientist to propose that species

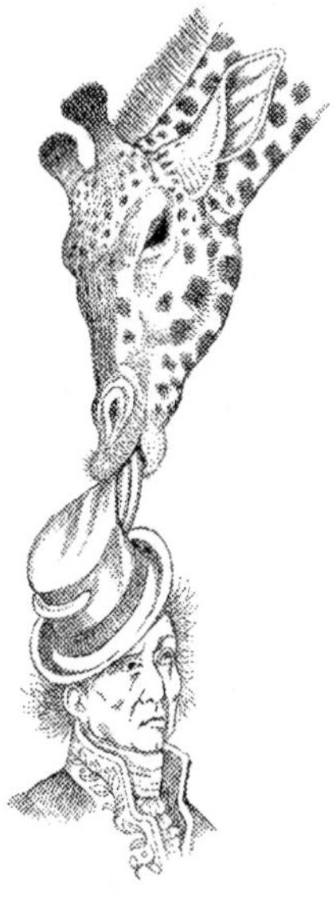

actually changed gradually over time and that they did so because of natural processes (as opposed to supernatural ones). In his spare time, Lamarck was the first naturalist to separate crustaceans, arachnids, and annelids from insects (although housecleaners had been coping with that very problem for many years), and he also coined the term *invertebrate.* All in all, a fairly hefty set of accomplishments—most of which go completely unmentioned, unappreciated, and most important (for high school students, at least), unmemorized. Instead, Lamarck has been hammered in nearly every introductory biology text ever written. Indeed Lamarck's folly, referred to as "the inheritance of acquired characteristics" has hung around the poor man's neck like an albatross (or more accurately, a giraffe).

In Lamarck's giraffe story, used to explain how evolution might proceed, there was once a population of short-necked animals (let's call them protogiraffes) feeding happily on low-lying leaves. For whatever the reason, the environment changed, these plants died out, and the short-necked animals were left with a dwindling food supply.* According to Lamarck, the protogiraffes that had previously fed on the vertically challenged (and now extinct) foliage, *needed* longer necks in order to feed from the higher branches of trees that hadn't been wiped out. This need somehow produced elongated necks, resulting in the evolution of the modern giraffe.

Although it took a bit of time to discredit Lamarck (Charles Darwin actually fell back on Lamarckian concepts in his later editions of *Origin of Species*), eventually folks came up with questions like "If Lamarck was right, then why are boys with circumcised fathers born with a foreskin?"†

*Long before anyone else figured it out, Lamarck actually nailed down the fact that an environmental change like the one just described was necessary in order for evolution to occur.

†Much later, a group of couch-bound scientists came up with an equally tricky question: "Why aren't the children of professional bowlers born with super bowling skills?" Presumably this question arose in the mid-1960s, when pro bowling became a wildly

In reality, most of what you do or experience in your lifetime has little or no effect on the genetic makeup of your offspring. Whether we're talking about longer snouts and legs for Miocene protohorses, or sharper teeth for ancient vampire bats, any inheritable modifications inevitably came about as a result of changes that occurred at the genetic level (i.e., changes in part of a genetic blueprint or in the timing of genetically programmed events).*

It's this change in the timing of genetically programmed events that explains how ticks may have evolved from chiggers. In this case, blood feeding may not have been a novel characteristic (as in ancient vampire bats), but possibly the timing of its appearance was. In a process known as heterochrony, the timing of developmental events is altered. Heterochrony, then, could explain the origin of the first tick from an ancestral mite—a mite that somehow maintained its larval feeding behavior into adulthood.

How could something like that come about?

There are numerous examples of this process in nature, but the most well known is neoteny—in which an organism reaches sexual maturity while retaining juvenile characteristics. The classic example concerns the giant salamander *Necturus* (the mudpuppy), which retains its gills throughout adulthood. In the vast majority of amphibians (like most salamanders, as well as mudpuppy cousins, like frogs and toads), these respiratory structures are lost as the larvae metamorphose into semiterrestrial adults.

It's been hypothesized that in this famous case of neoteny, a mutation allowed some salamanders to retain their gills as they

---

popular televised sport—a phenomenon in itself that has troubled many evolutionary biologists.

*And just as important, only if those changes occurred in that individual's gametes (their sperm or egg cells). Somatic mutations (like cancer), which occur in nonsex cells (like those of the skin), are not passed on to the next generation, although inherited mutations in a person's DNA could predispose him or her to develop skin cancer.

reached sexual maturity. The obvious question is, Why would that particular characteristic become an adaptation? The best hypothesis thus far is that the selection pressure to retain gills as an adult might have been a change in the terrestrial environment (e.g., a new predator or drier conditions), making it safer to extend the time salamanders spent in the ponds where they swam as larvae.

Similarly, with regard to the evolution of ticks, perhaps increases in local vertebrate populations or species diversity (both are also forms of environmental change) led to an evolutionary advantage for some mites that accidentally retained the parasitic dietary habits they had as larvae. Basically, more vertebrates meant more exploitable sources of food. As in *Necturus,* this adaptation had evolved from a mutation that hadn't produced a *new* character but instead had changed the developmental timing of a previously existing character. Following this hypothetical scenario to its conclusion, true ticks would have evolved as prototicks transitioned from feeding on liquefied cell contents (like their mite ancestors) to feeding on blood.

However ticks came about, most researchers think that the first ticks appeared sometime during the early Cretaceous period (around one hundred million years ago) and, not coincidentally, during a period of tremendous vertebrate diversity.

Within the arachnids, chiggers, ticks, and mites belong to the order Acari (or Acarina), which contains between 850 and 900 species of ticks and approximately 50,000 species of mites.

According to Gwilym O. Evans, author of *Principles of Acarology*, acarines are unlike other arachnids because of the intimate associations they've developed with other animals. In mites, these associations range from symbiosis to commensalism to parasitism.

Briefly, symbiotic relationships are those between two different organisms in which both derive some benefit. Among the acarids, perhaps the strangest example of symbiosis is the relationship between the eastern subterranean termite *Reticulitermes flavipes* and the slime mite *Histiostoma*. Researchers have found that termite colonies often become infected with a pathogenic fungus *(Metarhizium anisopliae)*. The fungus invades the termite's body, secretes a fatal toxin, and then derives nutrients from the decomposing wood muncher. Finally, the rootlike fungal mycelia erupt through the cadaver's exoskeleton to grow and spread reproductive spores throughout the termite colony. So destructive is this fungus that it's even been considered for use in the biological control of termites. Fortunately for the termite (although unfortunately for homeowners and pest-control types), the slime mites living in the nest not only scarf down the pathogenic fungus, but as they cruise around the nest they spread a trail of bacteria, yeast, and other microbial organisms. This sets up competition between the pathogenic fungus and these nonlethal decomposers with the result being the suppression of growth and sporulation (release of the reproductive spores) in *Metarhizium.* In many ways, it's as if the slime mite is able to serve as an external immune system for the termite.

Commensalism (another type of mite/animal association) is a relationship between two organisms in which one benefits and the other neither benefits from the relationship nor is harmed. One example, in the case of mites, is a form of commensalism known as phoresy, in which a smaller organism (in this case, the mite) attaches itself to other organisms (like an insect) for the purpose of transportation. Since the carrier isn't harmed, you can think of phoresy as a milder version of the passive transport we saw in bed bugs. In perhaps the strangest case of phoresy, hummingbird-flower mites *(Proctolaelaps kirmsei)* are chauffeured from flower to flower within the nasal cavities of the hummingbirds. Although the hummingbirds aren't physically harmed by the mites, they both wind up competing for the same pollen and nectar—and so this isn't really a textbook example of commensalism.

Acarologist Tyler Woolley lists five significant ways that mites affect humans: *health* (through transmission of diseases as well as our bodies' allergic and inflammatory reactions to them), *agriculture* (they infest crops, household and garden plants, and farm animals), *stored agricultural products* (they cause tremendous damage to grains, cereals, and veggies in which they live and multiply),*

*According to entomologists R. Chapman and H. Shepard, "Mites infest foodstuffs to such an extent that the entire mass may appear to be in motion . . . If some of the flour which is suspected of containing mites is piled in the light, the mites will crawl away from the light and the pile of flour will usually flatten out." And in a quote that immediately reminded me of Lou Sorkin and his citronella-scented bed bug colony, the authors stated that "when present in large numbers they give off a sweetish, musty odor, which is so characteristic that one who has had experience can detect their presence without having seen them."

biological control (in which predatory mites are involved in controlling pests like fire ants or even other mites), and *aesthetics* (nobody likes a mangy mutt or mite-damaged houseplants).

As a group, mites exhibit a bewildering variety of ways to make a living. For example, approximately 140 species of them have been identified as living in house dust. Additionally, if you look closely enough you'll find mites infesting algae, books, cheese, dried fruits, dried meats, drugs, flour, fungi, furniture, grains (like corn, wheat, oats, barley, rye, buckwheat, and millet), jams, jellies, mattresses, mildew, mushrooms, nectar, nuts, paper, plant bulbs, pollen, seaweed, seeds, spores, straw, sugar, vanilla pods, and wallpaper. Mites affect hundreds of plant species, and pretty much every type of wild animal, farm animal, and pet you can name. For creatures troubled by mites, infestation sites range from ears to anuses and all stops in between.

Besides an allergic reaction to dust mites and their droppings, perhaps the most commonly encountered mite-related health problem is scabies. Caused by *Sarcoptes scabiei.* Scabies is a condition that produces a rash and intense itching.* The symptoms result from the host body's reaction to mite-secreted and -excreted substances released as the mites go about their parasitic business. Young female scabies mites, which are about one-fiftieth of an inch long (a half millimeter), excavate a burrow in the host's skin where a male soon joins them. Copulation occurs only once and renders the female fertile for life. Soon after, she emerges from the honeymoon suite (leaving the male behind to die). The pregnant female motors around the surface of the host (reaching speeds of up to 60 inches per hour) until she locates a site for a permanent burrow (hands and wrists are popular). Burrowing at a rate of about one-fifth of an inch per day (five millimeters), the female feeds on liquid from ruptured host cells. She also takes

*The word *scabies* comes from the Latin *scabere,* "to scratch."

time to pump out several eggs per day, which are applied to the walls of the ever-lengthening burrow. When the larvae hatch, they leave mom and their nursery burrow behind, passing through several instars before reaching adulthood. During their wanderings topside, scabies mites are commonly spread to new hosts during periods of prolonged physical contact.

Until relatively recently, scabies was thought to be a disease of the poor, the unwashed, and the sexually promiscuous. This view was challenged in a rather unique manner in an article titled "Scabies Among the Well-to-Do," published in 1936 in the prestigious *Journal of the American Medical Association:*

> Scabies is a disease of herding, promiscuity and travel, of family school and vacation life. A plague of armies, tenements and slums. It may with equal force invade a pedigreed school, Camp Wawa Wawa or the baronial castle on the hill. An ever present differential consideration, wholly without social boundaries, the possible explanation of the itches of the tycoon, the socialite and the university professor equally with the mechanic's daughter on relief.

Another mite causing *major* concern today is *Varroa destructor,* which preys on several types of bees, including honey bees *(Apis)* and bumble bees *(Bombus). Varroa* can be considered an invertebrate vampire because it feeds on hemolymph. Since the bee's circulatory system doesn't function in gas transport, there is no oxygen-carrying hemoglobin, and as a result hemolymph lacks the red color of vertebrate blood. It is, however, a complex liquid containing a variety of hemocytes, cells that carry out many of the same functions as their leukocyte counterparts—functions that include phagocytosis and a role in the immune response. There's even a hemocytic version of stem cells.

Female mites enter bee nests (or hives) where they lay their

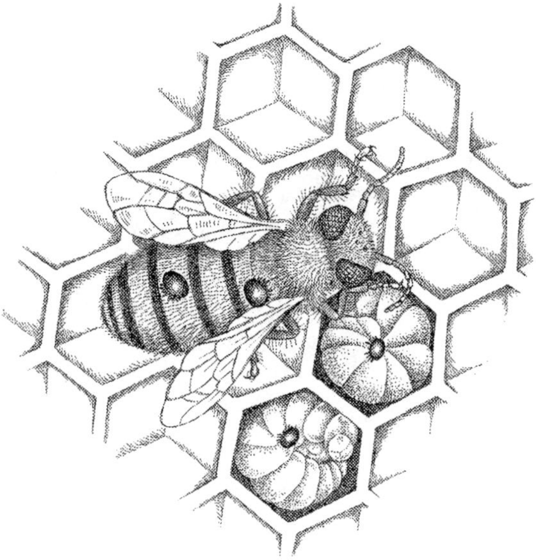

eggs just before the brood chambers containing the developing bees are capped by the adult bees. The parasites feed on larval and pupal instars as well as the emerging adult bees, which are also used for transportation. As with other arthropod parasites, as *Varroa destructor* feeds it can transmit viral and bacterial pathogens to its host.

Recently, the dramatic and nearly worldwide loss of honey bees has become a major concern not only within the beekeeping industry but also among farmers who raise the more than ninety commercial crops commonly pollinated by bees.* Colony collapse disorder (CCD, formerly known as fall dwindle disease) is characterized by the sudden departure of most of the adult worker bees from the hive, leaving behind the queen, a few young workers, and an abandoned brood of larvae and pupae. Although the cause of CCD is still under investigation, the list of potential suspects

*These include apples, pears, blueberries, almonds, pumpkins, and squash.

includes mites, bacteria, fungi, viruses, long-term exposure to substances like pesticides—especially neonicotinoids (chemicals that mimic the neurotoxic effects of the compound found in tobacco), and poor nutrition.* There is even a suggestion, albeit far-fetched, that cell phones are the causative agent.

In a pilot study published by the International Association of Agriculture Students, researchers at the University of Koblenz/Landau in Germany, placed cell phone handsets near four of eight beehives. They set out to measure hive-building behavior (by comparing before and after photographs of the hive chambers) as well as the tendency of the bees to return to their hives after they'd been captured, marked, and released some eight hundred meters away. Although the researchers reported that during the experiment "it became clear that both weight and area (of the hive) were developed better by non-exposed bees" statistical analysis "never showed a difference between exposed and non-exposed colonies." Oddly, in their Results section, the authors presented only *half* of their bee return data. They reported that in one exposed colony, only six of twenty-five test bees returned home within forty-five minutes, while in a second exposed colony, no bees returned. These incomplete findings triggered the publication of several articles (e.g., "Scientists Claim Radiation from Handsets Are to Blame for Mysterious 'Colony Collapse' of Bees," "Cell Phone Plague Obliterates Bee Colony," "Honey Bees Can't Call Home") purporting to inform readers of the dramatic new scientific developments. Typical was an editorial in the *Waco Tribune Herald* (April 16, 2007) in which the author stated that "a

*The malnutrition hypothesis posits that bees forced to pollinate large monoculture farms are missing something in their diets in much the same way that a dog fed nothing but bread would experience physical harm from such a diet and eventually starve to death. On a related note, weather (e.g., drought) can also negatively affect pollen-producing plants, resulting in pollen that is deficient in the nutrients the bees require.

growing theory is that cell phones cause bees to become so disoriented that they cannot find their way back home."

The original researchers were clearly not amused. According to Dr. Wolfgang Harst, the lead author, "This evolved as a case study for us in the new 'copy and paste' journalism." Harst slammed "the erroneous depiction of our study," from "faulty facts" about the study itself, to the claim that "handsets are to blame for 'colony collapse.' " He informed me that the follow-up study is set for publication in the journal *Environmental Systems Research* and that "although the findings are not so 'alarming' or 'breathtaking' as in 2005, the differences we found between the full irradiated and non-exposed bees were significant."

A number of researchers have published studies strongly suggesting that CCD is caused instead by a virus transmitted to bees (and/or activated) by *Varroa destructor*, the previously mentioned, hemolymph-sucking bee parasite.

Two closely related viruses have been implicated: Kashmir bee virus and Israeli acute paralysis virus.* These viruses are thought to be common infective agents within bee colonies (approximately eighteen bee viruses have been described) until stress or another problem (like *Varroa*) causes them to become epidemic and lethal.

"They've been selectively breeding different honey bee strains for years—for traits like mild temper, honey production, and resistance to mites," said Kim Grant, biologist and beekeeper. "It's certainly possible they've also bred in some things they *hadn't* planned on—like susceptibility to some of these bee viruses or compromised immune systems."

Currently, scientists are trying to determine methods to stop the

*Researchers have identified both of these viruses in nearly all hives with CCD but not in control hives.

spread of CCD—many of which involve *Varroa.* These include the development of new miticides and the introduction of *Varroa*-resistant bees into European and American bee colonies. Clearly, though, beekeepers and farmers are taking CCD extremely seriously since the potential exists for a global nightmare should the world's bee populations disappear.

Scientist and *New York Times* best-selling author Dr. Charles Pellegrino, a polymath whose novel *Dust* took an apocalyptic view of what would happen should the earth's insects go extinct, was less than optimistic about the ramifications of a honey bee extinction event.

"So what do you think is causing this?" I asked him in the spring of 2007, as we sat on my favorite bench in Washington Square Park.

"The feeling from people I've talked to with the CDC is that weakened bee immune systems seem to be the issue here, with mite infestations more of a secondary symptom."

"What's compromising their immune systems—cell phones?"

There was a pause and Dr. Pellegrino frowned. "You're kidding me, right?"

I shrugged.

"Well, it's still a bit of a poser," he continued. "If it's a viral agent like they're saying—even something akin to 'bee AIDS'—then I'm not terribly worried. Viruses usually adapt very quickly to their hosts—and a bad parasite usually ends up dead, inside its dead host. A viral problem can be expected to quickly self-correct."

"You mean evolve into a nonlethal strain?"

"Right. But if it's a fungus weakening their immune systems, that could be much more problematic."

"Why's that?"

"Fungi adapt more slowly than viruses or bacteria. Plus they're resistant to all but the sorts of antimicrobial agents that would kill the bees as well as their parasites."

I figured it was time to bring out the big guns. "What would happen if all the bees went extinct because of CCD?"

Dr. Pellegrino gave a chuckle, but there was no humor in it. "It doesn't need to be a total extinction event. If bee deaths should reach 80 to 90 percent worldwide, I estimate that the earth's carrying capacity for human beings could be reduced, essentially overnight, from a maximum of twelve billion to about six billion—and we're at six point seven billion now."

"So you think the result would be . . . ?"

"I think the result would be widespread famine and economic collapse, on a planet where the kamikaze mentality has already turned religious extremists into tigers sharpening plutonium claws."

"Okay . . . but why the huge effect—the lack of bee-pollinated crops?"

"That's part of it, Bill. We'd be reduced to harvesting wind-pollinated crops like wheat and corn. But just as important, certain organisms are keystone species—basically nature's cascade points. Should they go suddenly extinct, or should their numbers be greatly reduced, then the entire system is affected. The honey bee is one of those keystones. Knock them down, near to extinction, and our civilization is gone in five years. Without the honey bee, Rome falls."

We sat there silently for a minute, watching the chess players clustered at tables near the park's southwestern entrance.

"Checkmate," I muttered.

Pellegrino gave another humorless laugh. "You got that right."*

*In a development eerily similar to CCD, White Nose Syndrome has been killing thousands of hibernating bats in upstate New York and Vermont. The most obvious symptom is a white fungus around the nose of stricken and dead bats. Experts suspect that the fungus may be a secondary problem and that something else is killing the animals. In a press release from the New York Department of Environmental Conservation, bat specialist Alan Hicks said, "What we've seen so far is unprecedented. Most bat researchers would agree that this is the gravest threat to bats they have ever seen. We have bat researchers, laboratories, and caving groups across the country working to

Approximately one in three mite species belong to the suborder Prostigmata, and these are commonly known as harvest mites or scrub mites. Many of these species are relatively harmless as adults (feeding primarily on plant material) and some are actually beneficial—aiding in the decomposition of plant matter into humus—a vital soil component for the growth of plants. The problem is that somewhere between 2,500 and 3,000 prostigmatids (most belonging to the family Trombiculidae) have parasitic larval instars commonly known as chiggers.

Considering the grief that they cause, only a few species of chiggers count humans as their primary hosts. In that regard, most chigger/human encounters are accidental and generally end badly for both parties. Instead, a significant majority of chiggers parasitize nonhuman hosts, including many invertebrates (like arthropods) as well as every major group of vertebrates.

Chiggers, like their tick cousins, have a worldwide distribution, which means that you're just as likely to get bitten by a tick in Central Park as you are a chigger in Tunapuna, Trinidad, and although there are several species of chiggers in the United States (belonging to the genus *Trombicula*), the most commonly en-

---

understand the cause of the problem and ways to contain it. Until we know more, we are asking people to stay away from known bat caves." The DEC statement went on to say that, "Bat populations are particularly vulnerable during hibernation, as they congregate in large numbers in caves—in clusters of 300 per square foot in some locations—making them susceptible to disturbance or disease." Adding to the problem, most of the bats known to hibernate in New York do so in just five caves and mines. "We have lost more than 90 percent of the animals at the two sites for which we have good survey data," Hicks told me. "The problem is also expanding into new sites and now involves hibernacula harboring over 200,000 animals." Bat biologist John Hermanson was also deeply concerned about what was taking place. "We've got caves with eight bats in them. There used to be thousands."

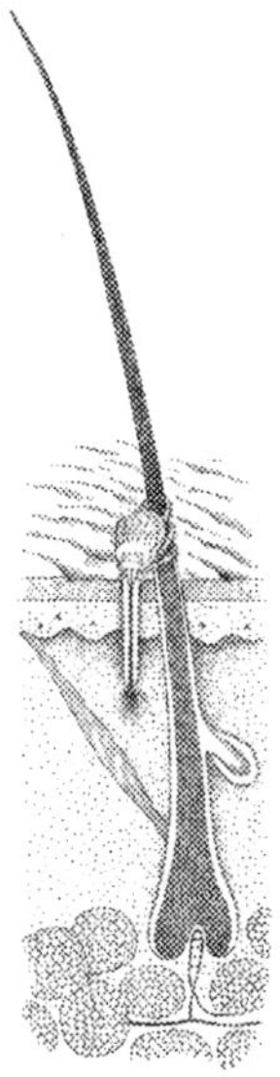

countered is *Trombicula alfreddugesi* (while in England it would be *Trombicula automnalis* and so on).

Although chiggers and ticks do exhibit some similarities, the differences between them are significant enough that they should not be confused with each other.

Besides diet, one major difference between chiggers and ticks is size. Chiggers are nearly impossible to see with the unaided human eye unless they're clustered together (most are about four-tenths of a millimeter long, which is about one one-hundredth of an inch). Ticks, on the other hand, can be hundreds of times larger.

Unable to hop aboard their prey like the long-jumping fleas,*

*Fleas are blood-feeding insects belonging to the order Siphonaptera (which numbers around 2,100 species). Like the chiggers that normally prey on rodents, but transmitted pathogenic bacteria to troops in Asia and the Pacific, fleas preying on rats transmitted the Black Plague to fourteenth-century Europeans. Following trade routes and

both chiggers and ticks locate their hosts either by actively hunting for them or by lying in ambush and waiting for them to brush past.

What occurs next—bite preparation, the bite itself, and the actual mechanism of feeding—is another area where chiggers and ticks differ significantly.

When attacking humans, chiggers move rapidly to areas of the body where the skin is particularly thin, like the ankles, armpits, or the back of the knees. Unlike ticks, they are remarkably fast runners, although both use a similar combination of sensory stimuli (light, touch, and chemicals) to track their potential prey. Upon encountering regions of the body bound by tight-fitting clothing (like socks, belts, or the elastic bands found on underpants and bras), instead of traveling over the material, chiggers crawl under it—often choosing these areas to initiate their bites.

One of the misconceptions about chiggers is that they burrow deeply into the skin of their hosts where they embed themselves (like ticks), but this is where the two parasites couldn't be more different. Once chiggers find a suitable patch of skin (usually an epidermal pore or at the base of a hair shaft), they pierce the skin with a pair of short fangs called chelicerae. As these daggers work back and forth, muscular contractions inject saliva into the wound. This saliva contains strong digestive enzymes that produce two very different reactions in the areas adjacent to the bite. Within a few hours, the outer layer of the epidermis immediately

---

the humans that plied them, black rats *(Rattus rattus)* spread across the world. As rat populations exploded, the fleas they carried began to encounter (and bite) humans on a regular basis. The plague struck with wavelike regularity throughout the Dark Ages, spreading inland from major ports, where it devastated cities and erased the inhabitants of entire towns. With no cure, the Black Death wiped out a significant portion of the human race (with estimates running as high as seventy-five million people killed). What saved civilization, apparently, according to one hypothesis, was that black rats were eventually replaced by another species, the brown rat *(Rattus norvegicus),* whose fleas were less likely to bite humans.

surrounding the injection site responds to the corrosive spit by hardening into a strawlike structure called a stylostome (or histiosiphon). The stylostome, which soon extends down into the dermal layer, is formed (at least in part) by keratin, a waterproofing substance released from the hosts' own epidermal cells. As the chigger's saliva flows down the stylostome's central canal, the powerful enzymes within it reach the deeper layers of the epidermis and eventually spread into the dermis. Here the enzymes liquefy the surrounding connective tissue and the contents of nearby cells. This cellular soup is the preferred diet of chiggers and although blood cells may accidentally become part of the recipe, they are not true vampires. The rudest part of the chigger's feeding gig begins as the liquefied dermal stew is snorked up through the stylostome and into the parasite's muscular pharynx.

Since chiggers generally feed continuously for three or four days, humans usually scratch them off long before they're finished. Once displaced, chiggers cannot attempt to feed again and die without developing further.

Meanwhile, back at the bite site, the host's immune system reacts to the stylostome and foreign chemicals. The resulting inflammation produces some serious and prolonged itching, which can lead to secondary infection.

In addition to misconceptions about how chiggers feed, another myth concerns ridding yourself of the pests (or at least alleviating the itching they cause) by applying clear nail polish to the irritated skin. The truth of the matter is that the chigger has probably been scratched off already and nail polish has never been renowned for its therapeutic properties. Instead, it's recommended that the bite area be cleansed thoroughly. Following this, antihistamines and topical anesthetics can help to alleviate the itching, but even so, the welts and the urge to scratch the bites can sometimes continue for ten days or longer—basically until the stylostome has broken down and is reabsorbed by the body.

Although most chiggers do not get to complete their human meals, some chiggers lucky enough to find a nonhuman host eventually drop off (generally within three days) and burrow into the ground. There, they go through two more larval stages before a final molt results in an eight-legged adult mite.

Ticks are a much smaller group than mites (or even chiggers) and they're divided into two superfamilies: the Ixodoidea (or hard ticks) and the Argasoidea (or soft ticks). Ticks are obligate blood feeders, and as such, their feeding habits are far more specialized than those seen in mites. Ticks feed solely on vertebrate blood, and they parasitize mammals, birds, reptiles, and amphibians, which basically means that they pester every major group except fish. Ticks are a huge problem for humans even though we are not the primary hosts of a single tick species.

Hard-bodied ticks, ixodids (which cause the most grief for humans), range from 1.7 to 6.1 millimeters in length and soft-bodied ticks can get even larger yet (3.6 to 12.7 millimeters). Amazingly, when their bodies are bloated with blood, ticks from both groups can reach lengths of between 20 and 30 millimeters.

In the United States and elsewhere, tick control has focused on hard ticks since these parasites are responsible for the transmission of eleven different diseases to human hosts (which places them second only to mosquitoes in the variety of diseases that they transmit to us). There are approximately eighty species of hard ticks in the United States, twelve of which are problematic for humans. Of these, three species pose serious problems.

The black-legged tick, or deer tick *(Ixodes scapularis)*, is responsible for the transmission of three human diseases, including Lyme disease. Two less frequently observed afflictions are babesiosis (a malaria-like infection that attacks red blood cells), and human granulocytic ehrlichiosis (a bacterial infection that is analogous to anaplasmosis—a form of "tick fever" in cattle). The primary reservoir for the Lyme disease bacterium, *Borrelia burgdorferi,* is the white-footed mouse *(Peromyscus),* which is apparently not sickened by the infection. *Borrelia* is transferred to the ticks when they obtain a blood meal from the mouse, and Lyme disease can result when the infected tick transfers the bacterium to other animals, like deer, dogs, and humans.

The American dog tick (sometimes called the wood tick), *Dermacentor variabilis,* is the primary vector for Rocky Mountain spotted fever, a potentially fatal disease caused by the bacterium *Rickettsia rickettsii.* Named for the area where the disease was first diagnosed, and for the characteristic spotted rash that occurs in places like the palms and soles of the feet, Rocky Mountain spotted fever begins with flulike symptoms that worsen as blood vessel linings are attacked and major organ systems suffer the consequences.

Finally, the Lone Star tick *(Amblyomma americanum)* is so named for the distinctive, roughly star-shaped silver marking on the dorsal surface of the female's body (the male has white markings along the posterior edge of its body). The Lone Star is a major concern because it transmits a "Lyme disease–like illness." This is no surprise to those who study ticks since the bacterium they transmit *(Borrelia lonestari)* is closely related to *Borrelia burgdorferi,* the spirochete that causes Lyme disease. Research on the Lone Star tick has been stepping up recently, mainly because of the way the tick is sweeping into the northeastern United States. In fact, in many places (like Long Island) it is swiftly replacing the black-legged tick as the species most commonly encountered by humans. A more aggressive predator than the black-legged tick, the Lone Star actively tracks its potential host rather than waiting for it to pass by.

According to entomologist Tamson Yeh, the Lone Star tick presents additional problems for both integrated-pest-control specialists and the public.

"In the past, tick-free zones could be set up in parks and

playgrounds. We did this by cutting back on brush and building mulch or woodchip buffers between areas of woods and lawn. But since the Lone Star ticks are more mobile, they have no problem crossing these buffers."

Interviewing Dr. Yeh at her office in Riverhead, New York, I also learned that while the black-legged tick is primarily a forest dweller, the Lone Star prefers its environments hot and dry with plenty of open spaces.

"Considering the changes in local moisture patterns, it's a no-brainer that this particular tick is becoming prevalent in the northeastern United States."

I shifted in my seat. "By changing moisture patterns, are you referring to global warming?" Considering the media bombardment surrounding this catch phrase, I felt slightly uncomfortable now that I'd finally used it in a complete sentence.

Dr. Yeh hesitated. "Yes, global warming is a consideration—but it's more than that. When you cut down a wooded area and throw down a bunch of houses, lawns, and concrete, things are going to get hotter and drier. Humans are changing vegetation patterns, and the urban environments they're creating are just what the Lone Star tick thrives in. Plus, more people equals more contact with ticks."

Ticks exhibit significant variation in their hunting techniques and scientists have used these differences as a handy way to categorize them. The soft ticks (argasids or argasoids) are primarily "habitat ticks"; that is, they encounter their hosts in nests, burrows, caves, or other dwellings. Not only do these habitats provide birds, bats, and rodents with a safe place to avoid predators, raise young, and sleep, but they also provide stable microenvironments for hundreds of parasite species, including argasoid ticks.

On the other hand, the hard ticks are considered "field ticks" because that's generally where they attack their hosts. These are also the ticks we hear the most about, which is no surprise since they're the ones responsible for transmitting pathogens that cause Lyme disease and Rocky Mountain spotted fever.

Animals often pick up ticks and chiggers when they sit or lie down in an infested area. The parasites use a combination of visual, chemical, and tactile (touch) cues to close in on their victims. Additionally, both chiggers and ticks can catch a premeal ride when their potential prey brushes against the grass or leaves where the parasites have been clinging with their crablike limbs. This works in the following manner.

When an animal moves through high grass or brush, it causes physical disturbances to the environment as it tramples and displaces the soil and vegetation in its path. Ticks and chiggers take advantage of the vibrations produced during this process. Both show strong inclinations to climb upward—and the result is that they spend much of their hunting time perched at the tips of grass blades, sticks, and other objects located close to the ground. They also congregate along the outer margins of leaves belonging to weeds and other low-lying plants. When these substrates vibrate (owing to a physical disturbance like that produced by the movement of a nearby animal), both chiggers and ticks respond by lifting and waving their two front legs. This "questing" response increases the likelihood that the parasite's legs (which bear an array of Velcro-like spikes, hooks, and bristles) will make contact with a potential host as it passes by. Once the initial contact occurs, ticks employ all eight of their legs (six in chiggers) to haul themselves onto the unsuspecting lunch wagon.*

*Tick and chigger researchers actually collect their specimens by attaching a piece of rough fabric (like fleece or flannel) to a rodlike handle. These "flags," "drags," or "drag cloths" are then passed across the tops of tall grass or other low-lying vegetation in the hope that questing parasites will latch on.

Dr. Yeh pointed out a behavioral difference related to questing that seemed to spell even more trouble for those seeking to avoid encountering these parasites.

"Around here, black-legged ticks generally quest in the early morning, when it's not too hot and dry for them. Because Lone Star ticks *prefer* the heat, they quest in the afternoon. Unfortunately, that's when they're most likely to encounter people."

As far as speed goes, ticks are far slower than chiggers—slower moving, slower to bite, and slower to drop off after a bite. Additionally, unlike chiggers, which usually make a mad dash for your socks or belt line, once ticks have hitched a ride, they walk around the surface of their host's body (often for hours) using thermal and chemical cues to search for a suitable feeding site.

Some ticks, like those that prey on humans, aren't particular about *who* or *where* they bite, while others are highly specific. For example, the larval instars of *Rhipicephalus evertsi*, a tick that preys on bovids (cows and their cud-chewing relatives), prefer the ears of their host while the adults of this species show a strong preference for attaching themselves around the anus.

The cold tolerant tick *Dermacentor albipictus* thrives in the northern latitudes. In regions like western Canada, it's responsible for Winter Tick disease in large ungulates like moose. These hoofed giants become so heavily infested with ticks (sometimes numbering nearly two thousand per individual) that they spend much of their time grooming themselves and incessantly rubbing against trees. The resulting hair loss (up to 80 percent in some cases) gives the stricken moose a gray or even white appearance instead of their normal dark brown color and has given rise to the term *ghost moose*. These animals are often emaciated from blood loss and exposure, and since their feeding behavior is severely disrupted, they also exhibit loss of stored body fat.

Tick bites, and their behavior afterward, also vary significantly

from what is seen in chiggers. After tilting their body (somewhere between a forty-five- and sixty-degree angle), ticks use larger versions of the chelicerae found in mites and chiggers to scissor their way into a host's skin, where they embed themselves. Ticks also attach themselves to a feeding site with a rodlike structure called a hypostome. Unlike the chigger's feeding tube (stylostome), the hypostome of the tick is an actual body part through which blood is drawn from the host. Backward-pointing, hooklike projections on the hypostome prevent the tick from being easily removed. Additionally, in many tick species, the salivary glands produce a substance that literally cements the parasite to its host until it finishes feeding. Larval and nymphal ticks (which are often confused with chiggers) have smaller mouthparts and their bites are not as deep as those of the adult (which can penetrate past the epidermal and dermal regions of the skin into the underlying hypodermis).

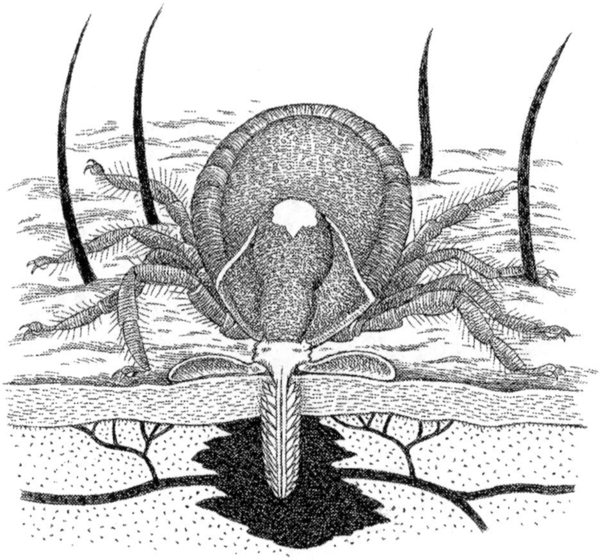

Deeply embedded into their hosts, ticks breathe through an insectlike tracheal system, with the openings (also called spiracles) located on a section of the abdomen that remains above the surface of the host's skin.*

As Dr. Yeh explained, there are a few unexpected positives related to the transition from black-legged to Lone Star ticks in the northeastern United States.

"First of all, the bite of the Lone Star tick is more painful," she said.

"That's always a big plus," I remarked, my mind wrestling with that one as I scribbled down notes.

"It makes them easier to detect," Dr. Yeh continued, unfazed

*Unlike insect respiratory systems, ticks have only two spiracular openings. Additionally, many ticks have a gill-like structure called a plastron that extracts oxygen from water, thus explaining why ticks can survive long periods of submergence. Although the mechanism is still being investigated, it appears that oxygen diffuses into the plastron and then into the tubelike tracheal system where it travels to supply the tick's body.

and exhibiting the patience that sometimes comes from working with the public.

"Right," I said, trying to figure out a selective advantage for a tick to have a *more* painful bite—there didn't seem to be any.

"Plus, STARI is milder than Lyme disease—although we don't know a whole lot about it yet."

"Starry?" I asked, trying not to sound like a bat biologist investigating tick-borne diseases.

"Southern tick associated rash illness. It's caused by *Borrelia lonestari,* the bacterium transmitted by the Lone Star tick. STARI has similar symptoms to Lyme disease—fatigue, flulike stuff, and a rash."

"What makes it milder?"

"It doesn't appear to go chronic—no long-term effects like Lyme disease can have on the joints, nervous system, and heart."

"Is there a test to distinguish between Lyme and STARI yet?"

The entomologist shook her head. "We haven't got Lyme disease figured out yet—and now we've got STARI to deal with."

"Why do the symptoms of Lyme vary so much?" I asked.

"If it isn't treated early—and sometimes even if it is—many researchers believe that the flulike symptoms can give way to a menu of far more serious stuff later on."*

"Is it tough to diagnose?"

"That's definitely one of the problems," Dr. Yeh continued. "The blood test for Lyme disease is notoriously unreliable, and the disease mimics other maladies—arthritis and multiple sclerosis. And

*Around 15 percent of patients who contract Lyme disease also report long-term neurological problems ranging from memory loss and diminished cognitive function to Bell's palsy (which is characterized by temporary facial paralysis), and even meningitis—a sometimes life-threatening infection of the protective tissues (the meninges) covering the outer surface of the spinal cord and brain. Meningitis can produce stiff necks, sensitivity to light, excruciating headaches, and far worse. The question of whether these long-term effects are actually caused by Lyme disease is a matter of current and heated debate.

apparently there's a latent form in which the bacteria that causes Lyme can hide in places like synovial fluid—where the host's immune system can't find it."*

As I interviewed Dr. Yeh, a debate was raging in the media and elsewhere among physicians, infectious disease researchers, bloggers, and various mothers-against-ticks groups over the existence (or nonexistence) of a chronic form of Lyme disease. Contrary to what Dr. Yeh believed, a growing contingent of physicians and infectious disease researchers find no evidence of a relationship between Lyme disease and long-term health effects. These professionals base their stance on recent studies that showed no trace of *Borrelia* in the spinal fluid, blood, and urine of well-documented Lyme disease patients who were acutely treated with antibiotics but complained of lingering symptoms months later.† Similarly, Dr. Lauren Krupp and her colleagues at Stony Brook University found that treatment of patients with long-term antibiotics produced results no better than with the placebo.

Those who opposed the concept of chronic Lyme disease offered several explanations for why patients previously treated for the disease showed serious health problems months and even years later. Supporters of the postinfection syndrome hypothesis proposed that *Borrelia burgdorferi* triggers neurological and other long-term problems in some Lyme disease patients *before* being wiped out by an acute course of antibiotics. Alternatively, some researchers felt that chronic Lyme disease patients who tested nega-

*Synovial fluid is found within the articular capsules that enclose joints like the knee, hip, and elbow. Bathing the cartilaginous surfaces where the bones meet, this slippery substance has the consistency of egg whites. It functions as a lubricant and also provides nutrients to the surrounding tissues of the joint. Apparently, some researchers (like Dr. Yeh) believe that synovial fluid also provides a safe haven for the corkscrew-shaped *Borrelia burgdorferi*, allowing it to elude the body's immune system (as well as administered antibiotics) and produce the chronic symptoms sometimes associated with Lyme disease.

†Most notable was a multicenter study led by Dr. Mark Klempner of Boston University School of Medicine.

tive for Lyme disease, or who had false positives, may never have had Lyme disease to begin with but were simply misdiagnosed.

The debate about chronic Lyme disease is still very much alive and there is abundant information out there from both sides of the fence.

I asked Dr. Yeh to clarify her position. "Do you think there are different strains of the pathogen?"

"There's every indication," she said. "It's one of the things that makes testing for Lyme disease so difficult. Too many false negatives."

I later learned that previous vaccination against Lyme disease, or prior treatment for syphilis or dental infections, could cause false positives by triggering an antibody response.

Suddenly, the withdrawal of the Lyme disease vaccine began to make sense.

"Is that what happened with Lymerix?" I asked.*

"That's one reason they stopped making it," Dr. Yeh responded. "The vaccine was designed to target a specific outer surface protein in *Borrelia burgdorferi,* but changes in those proteins have produced different bacterial strains. As a result, the vaccine was never all that effective to begin with—something like 40 percent. Add that to the fact that it was always a niche market, with maybe ten thousand doses in the entire United States. And then there was the big controversy a few years ago about vaccinations of all kinds. Should we give them? Are they safe?"†

"And it's not like vaccines are the big moneymakers," I chimed in, and Dr. Yeh nodded.

*Lymerix, the "breakthrough" vaccine for Lyme disease, was produced by the drug company GlaxoSmithKline and available between 1998 and 2002, when it was suddenly withdrawn from the market.

†On a related note, a recent lawsuit claims that the makers of Lymerix neglected to alert physicians and the public that around 30 percent of the population has a predisposition to an incurable form of autoimmune arthritis that can be triggered by the high concentrations of a specific bacterial surface protein contained in the vaccine.

Just as my interview with her was coming to a close, Dr. Yeh said something that sent a chill down my back. "Yeah, well, that moneymaking thing might change pretty quickly if bird flu ever evolves into something humans can pass off to each other."

And at that point, I realized that there was nothing left to say.

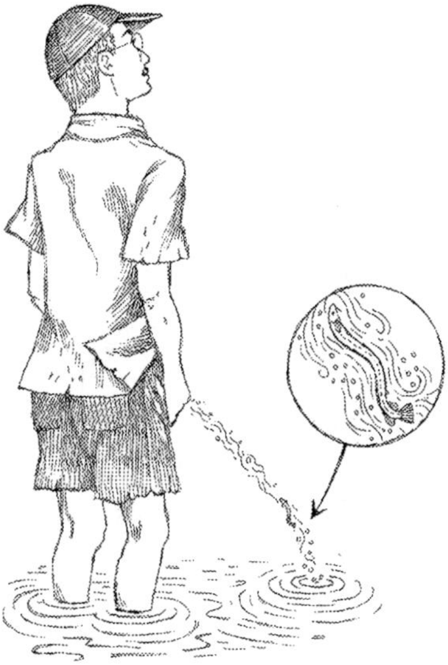

Very small, but solely preoccupied with doing harm.

—Paul Le Cointe

9.

CANDIRU

with a Capital *C* and That Rhymes with *P*

Several years ago I was involved in teaching a course at Long Island University that took twenty undergrads on a riverboat excursion up the Brazilian Amazon. We'd been in Manaus for three days, having spent the previous week at a rough-hewn research station called Kilometer 41,* and by the time we pulled into Manaus, we were all ready for a dose of civilization. The city was hot and crowded, but the food (and beer) were wonderful and the open-air markets

*My students were convinced that Kilometer 41 had been named for how far it was to the nearest flush toilet.

were fascinating—the bounty of the Amazon River on display in all its unrefrigerated glory. On the afternoon of our third day in Manaus, we checked out of our hotel and hauled our gear and belongings down to the harbor. There were boats of every size and shape lined up for what seemed likes miles, and I was reminded of the activity around a particularly busy anthill. Cargo was unloaded (all of it by hand) and replaced by throngs of people (all of whom seemed to be having a pretty good time). Fruit, fish, and vegetables of every size and shape were carried up a steep set of stairs that led to the city and its markets. Finally, we located the boat that was to be our home for two weeks. Gleaming white against the muddy and polluted water, the *Victoria Amazonica* was an eighty-foot riverboat with fourteen private cabins (each with showers and AC). In other words: we would no longer be roughing it.

Our captain was Moacir Fortes, the renowned Amazon guide. Mo was high octane, sharp as a stingray barb, and funnier than hell. After boarding the *Victoria Amazonica,* we stowed our gear and headed for a large centrally located room that would become a combination dining room/meeting place and indoor lounge. Captain Mo arrived shortly thereafter to address the group. He began by going through his famous list of "the seven perils of the Amazon River."

"There are thousands of creatures in the river that we enjoy—but only seven that enjoy us."

It was all very humorous as Captain Mo counted off some well-known local denizens—piranhas, stingrays, electric eels, anacondas, and black caiman (a South American cousin of the alligator). There was even a giant catfish, called the piraíba, that could swallow a person whole.

Finally, Captain Mo began describing a creature I hadn't heard of before. He was two sentences into his spiel when I noticed that the laughter in the room had stopped. In fact, the entire room had gone quiet.

The silence was broken several seconds later by the distinctive alarm call of a young North American woman: "It swims up your *what*?"

I was suddenly confused and I leaned in toward my friend, filmmaker Bob Adamo.

Bob was wearing a pained expression. He was also bent forward in his chair, as if he'd been punched in the gut.

"Did I just friggin' hear that?" I whispered.

"I hate it when fish swim up my Telegraph Office," came Bob's strained reply.

I had heard Mo correctly.

By 2006, as I began planning a return trip to the Brazilian Amazon, I had already learned quite a bit more about the candiru (or *carnero*), the creature whose very mention had caused such a commotion aboard the *Victoria Amazonica* five years earlier. I also knew that "Telegraph Office" was Captain Mo's unique way of referring to either human genitalia or the terminal opening of the digestive tract. In the case of the candiru, the term had been used because of the fish's legendary penchant for swimming up the human urethral opening and lodging itself there.

There was no shortage of candiru-related horror stories either, to the point where many sources claimed that it was even more feared by locals than its higher-profile river-mate, the piranha.* According to some of these authors, coconut shells, pudendal

*The *Victoria Amazonica* crewmember who lost a chunk of his thumb to a black piranha on our first trip to Brazil may have a slightly different opinion on the matter.

coverings made of dried palm leaves or bark, and wicker baskets (which might have served double duty on trips to the market) were worn to protect the external genitalia from candiru attacks.* But even these simple devices could be considered advanced anticandiru technology compared to the technique first described by early-nineteenth-century explorers Carl Friedrich von Martius and Johann Baptiste von Spix:

> These fishes are greatly attracted by the odor of urine. For this reason, those who dwell along the Amazon, when about to enter the stream, whose bays abound with this pest, tie a cord tightly around the prepuce and refrain from urinating.

Fortunately, as with the candiru's fellow sanguivore, the leech, there's a reliable and fairly comprehensive reference to this piscine vampire. In *Candiru: Life and Legend of the Bloodsucking Catfishes,* Stephen Spotte explores the bizarre world of these nasty creatures. Thankfully, the whole thing is done with clarity and some comic flair (i.e., it is not a textbook) and there's a nice chunk of science in it as well.†

Candiru belong to the Trichomycteridae, a family consisting of about two hundred species of "small to very small, slender-bodied freshwater catfish." Trichomycterids, most of which are rather plain-looking insect eaters, belong to the much larger group, the order Siluriformes (catfish), which consists of thirty-five families and around three thousand species. The catfish, in turn, fall under the incredibly broad heading of ray-finned fishes

*During a recent interview, candiru expert Dr. Stephen Spotte voiced extreme skepticism that this garb had anything to do with preventing candiru attacks and plenty to do with avoiding pests like ticks and sharp objects such as thorns.

†Dr. Spotte is a marine scientist and a prolific author He has published on topics ranging from modern zoos to mermaids. "I'm just an ologist," he told me.

(Actinopterygii), as do most fish you can name with the exception of sharks, skates, and rays.

Within the Trichomycteridae is a small subfamily, Vandelliinae, which currently contains six genera of obligate blood feeders inhabiting the Amazon and Orinoco rivers of South America. These vandelliines are commonly referred to as candiru (which Spotte pronounces "candy-roo"). According to Dr. Spotte, "Formal descriptions of candirus have been published since 1846, but the question of how many species exist has no clear answer." While some researchers think it's likely to be somewhere around fifteen (half of them belonging to the genus *Vandellia*), Spotte believes there could be far fewer. "Many species have been named on the basis of one specimen, and without the ability to measure variation within that species, it's impossible to determine if there's overlap with other supposed species."*

Physically, candirus are nowhere near as nasty looking as a decent-sized piranha. They are eel-like in appearance with their dorsal, anal, and pelvic fins set far back on their translucent bodies, close to the tail. Their small eyes are located near the top of their dorsoventrally flattened skulls. Candirus have tiny sensory barbells (unlike the prominent whiskerlike structures found in many catfishes), and they lack the stout and often dangerous dorsal and pectoral spines that have punctured countless anglers.

Although most people would consider this a good thing, the chances of ever seeing a candiru in the wild are quite low—that is, of course, unless you regularly fish the Amazon River and its tributaries with a cow lung tied to a piece of rope (a technique detailed in an article published by Kenneth Vinton and W. H.

*This would be comparable to aliens capturing a jockey and a basketball player, then calling them different species because of the observed size differences. Presumably, if they had a greater sample size (a whole town, perhaps), they'd realize that they were looking at a single species.

Strickler in 1941). The tiny fish usually remain hidden in the sand, mud, or leaf litter. They're also found in shallow, fast-moving water. This secretive lifestyle (which we've seen in other vampires) appears to be one of the reasons why studies on topics like the candiru's reproductive biology as well as many other aspects of its behavior remain sketchy or nonexistent.

Vendellia cirrhossa is probably the best-known species of candiru, although *Vendellia wieneri* has the coolest name. Candiru (which typically range from one to six inches in length) prey on larger fish and they feed by wiggling under their hosts' gill covers (opercula), the flaplike structures that shield the gill chambers. These opercula open and close as the fish breathes, and once inside, the tiny vampires secure themselves to the delicate gill lamellae (which are arranged like pages in a book) by a series of tiny, backward-facing hooks. These integumentary teeth or odontodes (sometimes called denticles) are found in patches around the candiru's head (including their own gill-covering opercular and interopercular bones). Once secured, the candiru utilize two or more rows of needlelike teeth to bite through one of the blood vessels that function in gas exchange between the feathery, high surface area gills and the fish's body.* Apparently, muscles in the candiru's mouth and pharynx pump the blood into the digestive tract.†

The candiru feeds for somewhere between thirty seconds and nearly three minutes, the blood clearly visible through its engorged, nearly transparent body. Since they sometimes feed in groups, the blood loss and damage to the host's gills can be considerable as pieces of shredded gill lamellae are sliced off by the teeth and rasped off by the odontodes.

Do candirus ever swim up the human urethra? Apparently, it does happen, although thankfully, occurrences are extremely rare. There have been numerous anecdotal descriptions, reviewed in depth by Spotte, but the first confirmed attack was reported by Drs. Spotte, Paulo Petry, and Anoar Samad at the 2001 meeting of

*In some ways this gas exchange mechanism is similar to the one that exists between the respiratory airways of terrestrial vertebrates and the tiny alveoli located in their lungs. Oxygen (which is at a higher concentration in the water surrounding the gills than it is inside the feathery gill filaments) diffuses from the water into these ultra-thin-walled structures. The oxygen then passes into blood being carried by tiny vessels within the filaments (in much the same way as oxygen passes from alveoli into the capillaries that surround them). These blood vessels then carry the oxygenated blood away from the gills to be distributed to the tissues of the fish's body. One major difference between the circulatory systems of fish and other vertebrates like amphibians, reptiles, birds, and mammals is that in fish, the oxygenated blood does *not* return to the heart before being pumped to the body. The fish heart is a rather simple, two-chambered pump (one atrium and one ventricle), rather than a three-chambered (amphibians and most reptiles) or four-chambered structure (crocodilians, birds, and mammals).

†A recent claim by researchers Jansen Zuanon and Ivan Sazima that candirus feed passively (i.e., the blood pressure of the prey literally pumps the blood into the candiru's digestive system) remains untested and Stephen Spotte is extremely skeptical. "From watching candirus feed, it appears that a rapid pumping mechanism of some sort is being used. Were ingestion passive, no such pumping activity, in fact, no movement would be necessary. The process would be like filling a bottle of water from a faucet."

the American Society of Herpetologists and Ichthyologists. Dr. Samad, a urologist, had treated a young man who showed up several days after a swim in the Amazon during which he had decided to remove his swim trunks before urinating.

> On 28 October 1997, one of us (Samad) attended a 23-year-old man who sought medical attention with extreme swelling and bleeding, having been attacked by a candiru. After extraction by endoscopy, the candiru measured 134 mm SL and 11.5 mm width across the head. Perfusion of the urethra with sterile distilled water prior to endoscopy induced immediate and pronounced scrotal edema. The candiru's penetration had been blocked by the sphincter separating the penile and bulbar urethras. Subsequently, the fish had bitten through the tissue into the corpus spongiosum, and the opening had allowed the perfusate to enter the scrotum. Some coagulated material was removed, revealing a wound on the bulbar urethra 1 cm in diameter and associated with a small amount of local bleeding. Because of its poorly preserved condition, the specimen could not be positively identified. However, it is likely that it is either a species of *Vandellia* or *Plectrochilus.*

According to a November 26, 2002, article in the *Niagara Falls Reporter,* by Pulitzer Prize winner John Hanchette, the man, identified only as FBC* was discharged after five days in the hospital, reportedly with no long-term effects from the encounter. The interesting thing, however, was that the patient's story had changed, igniting a firestorm of controversy that swept through candiru conspiracy buffs. FBC now claimed that he had *not* been submerged in the river when the attack occurred. Rather, he was

*The identity of the patient (Silvio Barbossa) was revealed in a more recent *Animal Planet* segment on parasites.

standing in thigh-high water when the candiru leaped from the river, darted through the urine stream, and lodged itself in his penis. After losing what can only be envisioned as a furious, but short-lived battle, FBC watched in horror as the five-and-a-half-inch candiru disappeared into this traumatized *chouriço.*

Could this happen? University of Calgary biomechanics expert Dr. John E. A. Bertram doesn't think so.

"In order to swim up a pee stream, the fish would have to swim faster than the stream flow. Additionally, to climb the stream, the candiru would have to lift itself out of the water against gravity. And while a small fish might be able to jump a surprising distance through the relatively low resistance air—a stream of urine would be another story."

"Why's that?" I queried.

"Because the penis tip acts as a very efficient nozzle," he said, "creating a stream of specific form, largely due to its high velocity of flow. Presumably, this keeps us from urinating on our feet. In any event, even if the candiru could power itself up the stream, it would have to stay completely within the urine and that would be difficult."

"How come?" I asked.

"Because the narrow urine stream has boundaries with the relatively low-density air. If the fish wandered into this boundary, the low resistance flow of the air would destabilize it—essentially pulling the candiru out of the high-drag pee stream."

I mentioned this explanation to Stephen Spotte, who agreed with Bertram. "I just don't see how it's possible," he said. "The candiru would be trying to swim, and the lateral sweeps of its tail would be wider than the urine stream. So in addition to destabilizing its body, it wouldn't be able to generate the thrust it needed."

"So was this Brazilian guy just making it up?"

"I don't think so," Spotte said. "I mean he didn't even know what candirus were, so it's hard to believe that he invented the story. I still think he was, you know, pissing right at the surface—and then it would be possible."

"How so?"

"If you've ever watched these things feeding in an aquarium, they shove themselves right under a fish gill—I mean it's a rapid, violent motion. It happens instantaneously—you can't even see it. So it's possible that this fish got up there so quickly that the guy didn't even have time to react. That part I believe."

"How long do you think a candiru could survive inside a human urethra?" I asked.

"Catfish can live a long time in pretty dire circumstances," Spotte replied, and I noticed that his voice had taken on an ominous tone.

"How long are we talking about?" I asked, a little fearfully. "A couple of minutes?"

"More like a couple of hours," he said, "although this one was certainly dead by the time they pulled it out of the victim."

I was almost afraid to ask the next question. "When was that?"

"They did the surgery three days after the guy came in. Besides the pain in his penis, I can't even imagine what it was like not being able to urinate for three days. Dr. Samad said that his abdomen was swollen like a soccer ball. He was a tough dude."

So why would a candiru abandon its normal, gill-feeding lifestyle for a visit to the Telegraph Office? In his book, Stephen Spotte reviewed several current hypotheses.

In the "urine-loving hypothesis," the candiru seeks to embed itself in a mammalian urethra with its ultimate destination being the urinary bladder.

"This whole idea of candirus being attracted to urine is problematic," he told me. "For one, urine is not listed among the major food groups." And in fact, no other vertebrates are known to feed solely or even predominantly on urine.

Then there was the oxygen—or lack of it actually. Oxygen is a requirement for every other vertebrate on the planet, and in a urine-filled bladder there wouldn't be any. Additionally, urine would be far saltier than the fresh water of the Amazon and the temperature would be significantly higher as well. All in all, it was a trip that sounded more suicidal (or accidental) the more I learned about it.

Interestingly, these apparent drawbacks didn't stop Carl Eigenmann, one of the top ichthyologists of his day, from proposing that "further study may demonstrate that some species of Candirús have become parasitic in the bladders of large fishes and aquatic mammals."

Once again, Stephen Spotte was not convinced. "Paulo Petri and I did some experiments, both in the field in sort of a quasi-laboratory situation, and we got no response at all."*

Another hypothesis proposes that candirus become accidentally embedded in the human urinary tract owing to a behavior called rheotropism, which is defined as the movement of an organism in response to a current. The idea here is that the candiru mistakes the flow of human urine for that of a natural flow of water (like the current produced as water leaves the gills of a large host fish). In what has become known as the "wrong turn hypothesis," after the candiru mistakenly enters the urethra, its backward-curved odontodes prevent it from turning around. The trapped creature continues forward until it dies from a lack of oxygen.

Yet another hypothesis posits that the candiru tracks its piscine prey by the chemical trail the larger fish leaves in its wake. If chemicals typically found in human urine (possibly ammonia, the protein albumin, and creatinine—a breakdown product of muscle physiology) also serve to stimulate the candiru's host-hunting

*In separate trials, they added solutions of urine, blood, and ammonia to tanks containing captive candiru, and in each case they saw no visible change in the candiru's behavior. In fact, only when they released a fish into the water did the candiru perk up.

behavior—this might explain the creature's attraction to the human urethra. Sweat has also been suggested as an attractant, although this doesn't explain why the candiru headed for the tip of FBC's penis.

"It could certainly be something in the urine," Spotte told me. "It could also be the urine flow itself."

I decided to try out my own hypothesis. "What about the possibility that the candiru are reacting to a disturbance in the water—at least initially. They think they're reacting to a large catfish, and by the time they get close enough, it's too late."

"The problem with that is the fact that Paulo and I caught many of these fish in the middle of the night, in rapids—I mean you couldn't even stand up in this water. And yet we tethered a catfish out there and within thirty minutes the candirus found it. Now they weren't detecting movement because this catfish wasn't swimming."

"So what do you think the candiru are detecting?" I asked.

"Nobody really knows, but my best guess would be that it's a combination of things. Catfish can taste and smell—they've got all these really well-defined sense organs, specialized for dealing with low-visibility environments. Probably, they sense something in the fishes' protective slime—which is sloughed off continuously, just like we drop old hairs. Candirus also have an array of pores around their heads, so it's also likely that they're detecting the electromagnetic charges generated by things like muscle contraction. In any event, it's an interesting problem and it just goes to show catfish are amazing creatures."

"What about attacks on nonhuman mammals," I asked, adding that I thought it strange not to have come across any in the literature.

"None," Spotte replied. "We haven't got any evidence. But you're right. Why don't they attack dolphins and manatees and river otters—they're certainly releasing stuff into the water—and they've got some very large orifices."

The scientist continued. "As far as human encounters with candirus, I just can't think of it as anything other than an accident. But how it arrives there—that's still a mystery."

I asked Dr. Spotte a final question. "What do you think the odds are that someone submerged in a stream where candiru live, and deciding to take a pee, would get attacked by these creatures?"

Dr. Spotte replied without hesitation. "About the same as being struck by lightning while simultaneously being eaten by a shark."

What was so thought provoking about all sorts of Galápagos finches to young Charles Darwin . . . was that they were behaving as best they could like a wide variety of much more specialized birds on the continents. He was still prepared to believe, if it turned out to make sense, that God Almighty had created all the creatures just as Darwin found them on his trip around the world. But his big brain had to wonder why the Creator in the case of the Galápagos Islands would have given every conceivable job for a small land bird to an often ill-adapted finch? What would have prevented the Creator, if he thought that the islands should have a woodpecker-type bird, from creating a real woodpecker? If he thought a vampire was a good idea, why didn't he give the job to a vampire bat instead of a finch, for heaven's sake? A vampire finch?

—Kurt Vonnegut

10.

A TOUGH WAY TO MAKE A LIVING

In addition to vampire bats, leeches, bed bugs, ticks, mites, and candirus, there are literally thousands of species that feed on blood. They range from intestinal nematodes called hookworms, that produce iron-deficiency anemia in their hosts,* to mosquitoes, one of the

*Interestingly, humans with hookworm infections are half as likely to have asthma or hay fever. The idea is that some parasites survive by down-regulating their host's immune system. With its primary defense mechanism throttled back, it's far less likely that the body will mount an inflammatory response against a harmless allergen or attack its own tissues. On a related note, it appears that in addition to selecting for more antibiotic resistant bacteria, our "99 percent germ-free" culture has resulted in hypersensitive immune systems. The end result is an increase in the incidence of asthma, allergies, and some autoimmune diseases.

eleven families of fly-relatives (dipterans) with blood-feeding members. There are assassin bugs (which may have plagued Charles Darwin) and vampire moths (seven species in the Asian genus *Calyptra*). These insects use their sharpened mouthparts to pierce the skin of animals like water buffalo, elephants, tapirs, and even humans.

Then there's *Geospiza difficilis,* the pint-sized vampire wannabe that so impressed author Kurt Vonnegut. Although dramatic, "vampire finch" is not a terribly accurate name for this bird, which is also known as the sharp-beaked ground finch. Basically, the vampire tag is questionable, because unlike the other sanguivores mentioned thus far, *Geospiza difficilis* is not an obligate blood feeder.* The birds merit mention here, however, because they appear to be giving part-time vampirism a serious shot. In that regard, they occasionally supplement their normal diet of small seeds and nectar by eating eggs and by pecking at the wings, body, and tail regions of another Galápagos resident—the blue-footed booby *(Sula nebouxi).* Once *Geospiza difficilis* has inflicted a small wound with its beak, it begins sipping the booby's blood, hopping out of the way if the larger bird gets annoyed. The finches feed in this manner for several minutes, giving way to other individuals who line up, waiting their turn like customers at a deli counter.

Geospiza difficilis is widely distributed throughout the Galápagos archipelago, but only two populations feed on blood. Researchers have noted differences in feeding behavior, size, and vocalizations in these birds—differences that could indicate that a

*Likewise, mosquitoes also feed on substances other than blood, namely nectar and fruit juice. In *Anopheles,* the genus that transmits malaria, only the female mosquito is obliged to seek out a blood meal. Unfortunately, the lack of card-carrying vampire status is completely irrelevant when one looks at the astronomical death toll associated with these insects. For example, it's been estimated that mosquito-transmitted malaria kills one person every twelve seconds.

new finch species is forming before their eyes. For my own part, it's just as interesting to imagine what might happen if *Geospiza difficilis* became a bit more adept at obtaining a blood meal, and a bit more dependent on finding one on a daily basis. What if conditions on the tiny islands of Darwin or Wolf changed, and the seeds and flowers these birds needed to survive were no longer available? Would the blood-sipping finches simply relocate to another island? Perhaps, but if they did, they'd certainly run into competition from the well-established finch species already living there. Maybe the vampire wannabes would die out or interbreed with non-blood-feeding members of their own species. Or maybe *Geospiza difficilis* would accumulate a few more beneficial mutations (to its digestive system perhaps), until it had evolved into a *real* vampire finch—a feathered version of *Desmodus rotundus*.

Beyond the hypothetical, though, blood feeding makes absolute sense—and because it does, it should come as no shock that there are thousands of obligate sanguivores out there, as well as others that supplement their diets with blood.

But why does a blood-feeding lifestyle make sense? To address that issue, I'll start off by repeating the two questions I hear most often about vampires: (1) Why do creatures like ticks, vampire bats, and bed bugs even exist? and (2) What would it matter to the earth if all the blood-feeding creatures suddenly disappeared?

These are questions that I've been hearing pretty much since I began working on vampire bats in 1990, and while the first question is certainly valid, the second question illustrates a basic problem that most people have with science. That is, they don't think like scientists. Don't get me wrong, though. It's hard not to empathize with anyone who's been swarmed by black flies, suffered through malaria or Lyme disease, or experienced the twitching paranoia of a bed bug infestation. For most people, then, it's only natural to envision a world where these nasty critters didn't pester, sicken, and kill us with such incredible efficiency.

But vampires, whether they're bats, leeches, or bed bugs, don't exist to sicken or kill us. They exist because their ancestors evolved certain characteristics that allowed them access to a highly specific but worldwide resource, a resource that they could utilize as food.* That resource, blood, has been vital to the existence of every vertebrate that has ever swum, crawled, walked, run, or flown.†

With the aid of a muscular pump, blood travels through a wildly complex system of interconnected tubes. Sanguivores have evolved ways to exploit the accessibility of these tubes and, just as important, the fact that they can be opened and tapped. Given the potential for running into creatures hauling around these tubular filling stations, it would be remarkable if sanguivory *hadn't* evolved in a diverse array of taxonomic groups. But here's the catch. No matter how different blood feeders are from each other (exemplified by leeches and bats), there appear to be a finite number of ways that a sanguivore can successfully gain access to a meal. For this reason, vampires as different as bats and leeches share separately evolved but similar adaptations for their highly specialized lifestyles—that is, they exhibit convergence.

All vampires, for example, are relatively small in size. The largest appears to be the common vampire bat, *Desmodus rotundus*, tipping the scales at just under an ounce and a half (around forty grams). The reasons for this size constraint are apparently related to my Trinidadian mentor Farouk Muradali's mantra, "Feeding on blood is a tough way to make a living." In that regard, the more of the red stuff a vampire requires each day (or night), the less likely

*The oldest evidence for a blood-feeding lifestyle comes from a fossil "protomosquito" from the Triassic period (approximately 220 million years ago). Since there were no flowering plants present, and therefore no nectar to sip, the elongated proboscis of this insect presumably functioned in much the same way as it does in extant mosquitoes.

†It also comes in a second flavor (hemolymph), which is shed in a variety of arthropods groups.

it is to obtain it. On a related note, larger vampires would need to drain greater volumes of blood from their hosts, which would increase the likelihood of weakening them to the point of death—a maladaptive trait for a parasite. Additionally, large vampires would be easier to detect by their prey as well as their predators, therefore tending to neutralize another characteristic that all vampires share: stealth. This ability to avoid detection is employed in a variety of ways by blood feeders, both during their approach to a potential meal and during the actual feeding.

The list of convergent characteristics goes on.

All blood feeders also possess finely tuned sensory systems. These allow creatures as different as bed bugs and vampire bats to efficiently locate their potential meals, often in the absence of light.

Additionally, once vampires have situated themselves within striking distance, they inflict relatively painless bites with an array of razor-sharp cutting instruments. These include denticles (leeches), chelicerae (mites, ticks, chiggers), needlelike stylets (mosquitoes and other insects), and actual teeth (vampire bats). The sharpened nature of these structures allows the vampires to gain access to the blood of their hosts without causing alarm, but even so, the complications encountered during a blood meal are *far* from over.

One major problem that all vampires must overcome is hemostasis, or blood clotting. This process actually consists of a maddeningly complex cascade of chemical reactions that must occur before a clot forms.* For the creature carrying around all that blood, the key benefit to this hemostatic complexity is that it prevents blood from clotting where and when it shouldn't. The downside to the clotting cascade is that it has enabled blood

*In a chemical cascade, the products of each step in the chemical reaction are integral to the step that follows.

feeders to interfere with the clotting process at multiple points along the chemical pathway. In other words, if there were only one step in the clotting process where the potential blood feeder could thwart the process, the odds of evolving that ability would be pretty low. But if blood clotting can be disrupted at any one of many points in a complex chemical cascade, the odds would be much higher that such a clot-disrupting substance would evolve. As a result, although each vampire has its own separately evolved anticlotting substances, the outcome is identical—freely flowing blood from the prey, with clot formation delayed until after (sometimes *long* after) a blood meal has been obtained.

Considering how long their "manufacturers" have been in business, these natural anticoagulants are often far more efficient than anything produced by man, and several of them have become important medications. For example, the clot-dissolving properties of the vampire bat–derived substance desmokinase have been used to combat strokes, while hirudin, an anticoagulant found in leech saliva, is used to prevent blood clots from forming after hip replacement surgery. Additional vampire-derived compounds (like anesthetics, perhaps) have tremendous potential for use in the field of medicine and we can certainly expect to hear more about them as researchers explore the field.

But besides a potential for providing us with some useful pharmacological products, do blood feeders provide any additional benefits?

The answer is most certainly yes.

Vampire bats remain great examples for learning about the pitfalls of scientific discovery, especially how problems can arise when prejudice and misinterpretation substitute for careful observation and experimentation. Unfortunately, some of these early errors have perpetuated a slew of misconceptions about bats (although the next time someone implies that most bats are blood feeders, you'll be ready to pounce). Each year, thousands upon thousands of beneficial bats are killed because of fear and ignorance, and the problems that do exist with vampire bats (generally, *Desmodus*) are actually the result of man's destruction of the natural environment, combined with our dogged insistence on propagating domestic animals in places where they just don't belong. *Desmodus* and its cousins, *Diaemus* and *Diphylla,* display an amazing array of adaptations for their blood-feeding lifestyles and as such they are poor representatives for bats as a group. Far more typically, bats help reduce the number of harmful insects, pollinate plants that are essential to their ecosystems (and to humans), and help to reforest tropical regions shattered by slash-and-burn agriculture.

On another very important level, sanguivores provide food for a variety of other creatures. Leeches, for example, are a preferred meal of many species of freshwater fishes, a fact hasn't gone unnoticed by the fishing bait industry (one company offers a discount for leech orders over twenty-five pounds). Fishermen use leeches to catch game fish such as walleye *(Sander vitreus),* smallmouth bass *(Micropterus dolomieu),* and northern pike *(Esox lucius),* as well as smaller pan fish like bluegills *(Lepomis macrochirus).* Mosquitoes are another important food source, especially for birds and bats.* Mites, ticks, and chiggers are also eaten, and even bed bugs are food for some insects, including several species of ants.

Although I'm unaware of any studies on this issue, it's likely that blood feeders also serve an additional ecological role: namely, to cull out the old and the sick from prey or host populations. A moose that is able to emerge alive from a winter's stint as a "ghost moose" might very well be carrying around a genetic blueprint with an amped-up emphasis on surviving harsh conditions

*There is some debate as to just how many mosquitoes an aerial insectivore like a little brown bat *(Myotis lucifugus)* can eat in a single night. Estimates range from an insubstantial number to six hundred mosquitoes per hour (a number that always seemed a bit high to me).

(especially since the ticks do not feed on the moose during the summer). On the other hand, moose that die from winter tick infestations apparently do so from starvation (distracted from feeding by all that grooming and rubbing). Hypothetically at least, not only might that moose be carrying "inferior" genes, but its death would leave more food for heartier individuals and those unaffected by the tick onslaught.

Rather than wishing for blood-feeding creatures to disappear, or starting to twitch at their very mention, we should be dealing with the fact that vampires *are* here and they're probably here to stay. In that regard, some of these blood-feeding creatures, like mosquitoes, can be deadly enemies and should be treated as such—although I am certainly not advocating the wholesale application of pesticides. Other sanguivores, like the common vampire bat, bed bugs, ticks, and chiggers, can become serious problems—some of them with the potential to sicken or even kill us. We should keep in mind, however, that in most cases these vampires would rather be feeding on something other than humans and it's generally *our fault* when we encounter them.

Then there are blood feeders that have a high gross-out quotient but are basically harmless (at least to humans). Leeches fall neatly into this category, as do candirus (except on *extremely* rare occasions).

Finally, there are some vampires that will certainly require our help if they are to avoid extinction over the next few decades. The bird-biting vampire bats *Diaemus* and *Diphylla* immediately come to mind. In my opinion, even if you're not a fan of these creatures, with only five thousand species of mammals, we should not stand by as two of them disappear forever. It should also be stressed that conservation measures shouldn't be limited to vertebrate blood feeders. As researchers like Mark Siddall have recently shown, there are invertebrate species, such as the misclassified leech *Hirudo verbana,* that appear to have squirmed through the cracks in our wildlife protection laws.

In the words of Edward O. Wilson:

> We should judge every scrap of biodiversity as priceless while we learn to use it and come to understand what it means to humanity. We should not knowingly allow any species or race to go extinct. And let us go beyond mere salvage to begin the restoration of natural environments, in order to enlarge wild populations and stanch the hemorrhaging of biological wealth. There can be no purpose more enspiriting than to begin the age of restoration, reweaving the wondrous diversity of life that still surrounds us.

The tragedy of extinction is that not only do organisms disappear before we know the answers to our questions, they sometimes disappear before we know the right questions to ask.

NOTES

1: Wallerfield

17 **In 1933 Greenhall and Raymond Ditmars** Raymond L. Ditmars and Arthur M. Greenhall, "The Vampire Bat—A Presentation of Undescribed Habits and Review of Its History," *Zoologica* 4 (1935): 53–76.

18 **Heavily loaded down after a blood meal** J. Scott Altenbach, *Locomotor Morphology of the Vampire Bat, Desmodus rotundus,* Special Pub. No. 6, American Society of Mammalogists (Lawrence, Kans.: 1979), 19–30.

19 **It had taken me six months** William A. Schutt Jr., John Hermanson, Young-Hui Chang, Dennis Cullinane, J. Scott Altenbach, Farouk Muradali, and John Bertram, "Functional Morphology of the Common Vampire Bat, *Desmodus rotundus,*" *Journal of Experimental Biology* 200, no. 23 (1977): 3003–12.

21 **In 1941 Captain Lloyd Gates** David E. Brown, *Vampiro—The Vampire Bat in Fact and Fantasy* (Silver City, N. Mex.: High-Lonesome Books, 1994), 77.

2: Children of the Night

35 **In 1801, in Paraguay** David E. Brown. *Vampiro—The Vampire Bat in Fact and Fantasy* (Silver City, N. Mex.: High-Lonesome Books, 1994), 15.

35 **The vampire bat is often the cause** Charles R. Darwin, *A Naturalist's Voyage* (London: John Murray, 1886), 22.

38 **This is an adaptation that comes in handy** Uwe Schmidt, "Orientation and Sensory Functions in *Desmodus rotundus,*" in *Natural*

History of Vampire Bats, ed. A. M. Greenhall and U. Schmidt, 150–52 (Boca Raton, Fl.: CRC Press, 1988).

38 **A recent study suggests that *Desmodus*** Udo Gröger and Lutz Wiegrebe, "Classification of Human Breathing Sounds by the Common Vampire Bat, *Desmodus rotundus, BMC Biology,* 4, no. 18 (2006): 1–8.

40 **For example, Muslim gypsies in the Balkans** Matthew Bunson, *The Vampire Encyclopedia* (New York: Gramercy, 1993), 218, 278.

41 **In Victorian England** Jerry Hopkins, *Extreme Cuisine* (North Clarendon, Vt.: Tuttle Publishing, 2004), 269.

42 **A boy by the name of Ernest Wicks** Montague Summers, *The Vampire: His Kith and Kin* (London: Kegan Paul, Trench Trubner and Co., 1928), 46.

43 **Vlad's favorite torture method** Radu Florescu and Raymond T. McNally, *Dracula—A Biography of Vlad the Impaler* (New York: Hawthorne Books, 1973), 76–77.

43 **How did a murderous Romanian prince** Ibid., 8–9.

49 **Another hypothesis on the origin** M. Brock Fenton, "Wounds and the Origin of Blood-Feeding in Bats, *Biological Journal of the Linnean Society* 47 (1992): 161–71.

52 **As Stephen J. Gould explained** Stephen Jay Gould, *Wonderful Life: The Burgess Shale and the Nature of History* (New York: W.W. Norton, 1989).

52 **As an alternative to previous speculation** William A. Schutt Jr., "The Chiropteran Hindlimb Morphology and the Origin of Blood Feeding in Bats, in *Bat Biology and Conservation,* ed. T. H. Kunz and P. Racey, 157–68. Washington, D.C.: Smithsonian Institute Press, 1998).

3: Snapple, Anyone?

63 **Had he dissected a specimen** Thomas H. Huxley, "On the Structure of the Stomach in Desmodus Rufus," *Proceedings of the Zoological Society of London* 35 (1865), 386–90.

63 **In an experiment using barium-laced cow blood** G. Clay Mitchell and James R. Tigner, "The Route of Ingested Blood in the Vampire Bat," *Journal of Mammalogy* 51, no. 4 (1970): 814–17.

64 **According to a 1962 paper** William A. Wimsatt and Anthony Geurriere, "Observations on the Feeding Capacities and Excretory Functions of Captive Vampire Bats," *Journal of Mammalogy* 43 (1962): 17–26.

64 **George Goodwin and Art Greenhall took** George Goodwin and Arthur M. Greenhall, "A Review of the Bats of Trinidad and Tobago," *Bulletin of the American Museum of Natural History* 122 (1961): 187–301.

66 **In 1969, Cornell vampire bat expert** William N. McFarland and William A. Wimsatt, "Renal Function and Its Relationship to the Ecology of the Vampire Bat, *Desmodus rotundus*," *Comparative Biochemistry and Physiology* 28 (1970): 985–1006.

74 **Isolated from a clover mold** Robert and Michèle Root-Bernstein, *Honey, Mud, Maggots, and Other Medical Marvels* (New York: Houghton Mifflin, 1997), 95.

74 **After capturing the bats in mist nets** Rexford Lord, "Control of Vampire Bats," in *Natural History of Vampire Bats*, ed. A. M. Greenhall and U. Schmidt, 217–20. (Boca Raton, Fl.: CRC Press, 1988).

75 **A related, but less-cost-efficient, method** Ibid., 219.

76 **Although we didn't realize it** Bill Hayes, *Five Quarts: A Personal and Natural History of Blood* (New York: Random House, 2005), 172–73.

76 **Some researchers use an alternative method** Janet M. Dickson and D. G. Green, "The Vampire Bat *(Desmodus rotundus)*: Improved Methods of Laboratory Care and Handling," *Laboratory Animals* 4 (1970): 40.

77 **During the three years that we maintained** William A. Schutt Jr., Farouk Muradali, Mondol, Keith Joseph, and Kim Brockmann, "The Behavior and Maintenance of Captive White-Winged Vampire Bats," *Diaemus youngi* (Phyllostomidae: Desmodontinae). *Journal of Mammalogy* 80, no. 1 (1999): 71–81.

82 **Another way that *Diaemus* differs** Arthur M. Greenhall, "Feeding Behavior," in *Natural History of Vampire Bats*, 123–35.

83 **In 1984, zoologist Gerry Wilkinson** Gerald S. Wilkinson, "Reciprocal Food Sharing in the Vampire Bat," *Nature* 308 (1984): 181.

86 **So named for the frill** Karl Koopman, "Systematics and Distribution," in *Natural History of Vampire Bats*, 7–17.

88 **If you examine the hind limb bones** William A. Schutt Jr., "Chiropteran Hindlimb Morphology and the Origin of Blood Feeding in Bats," in *Bat Biology and Conservation,* ed. T. H. Kunz and P. Racey, 157–68. (Washington, D.C.: Smithsonian Institute Press, 1998).

88 **Researchers in the 1970s** Dona Howell and J. Pylka, "Why Bats Hang Upside-Down: A Biomechanical Hypothesis," *Journal of Theoretical Biology* 69 (1977): 625–31.

89 **Many bats have a structure called a calcar** William A. Schutt Jr. and Nancy B. Simmons, "Morphology and Homology of the Chiropteran Calcar," *Journal of Mammalian Evolution* 5, no. 1 (1998): 1–32.

91 **Basically, what I'd proposed was similar** William A. Schutt Jr. and J. Scott Altenbach, "A Sixth Digit in *Diphylla ecaudata,* the Hairy-Legged Vampire Bat," *Mammalia* 61, no. 2 (1997): 280–85.

93 **Rather than feeding from below the branch** J. Moojen, "Sanguivorismo de *Diphylla ecaudata* Spix em *Gallus domesticus* (L.)," *O Campo* 10 (1939): 70.

4: Eighty Ounces

102 **"It has been my unvaried rule"** "The Death of George Washington, 1799," EyeWitness to History, 2001, http://www.eyewitnesstohistory.com.

102 **Soon after, the incision was made** George Washington: Eyewitness Account of His Death," 2003, http://www.doctorzebra.com/prez/z_x01death_lear_g.htm.

106 **In desperation, Dr. Dick** Oscar Reiss, *Medicine and the American Revolution* (Jefferson, N.C.: McFarland and Co., 1998), 234–35.

107–8 **Other suggestions included rubbing** Ibid., 235.

109 **For example, some ancient Egyptians** Henry E. Sigerist, *A History of Medicine,* vol. 1: *Primitive and Archaic Medicine* (New York: Oxford University Press, 1951), 247.

109 **The word *blood* shows up** Douglas Starr, *Blood: An Epic History of Medicine and Commerce* (New York: Knopf, 1998), xiv.

109 **Since the ancient Hebrews believed** Kenneth Walker, *The Story of Blood* (New York: Philosophical Library, 1962), 20–22.

114 **Galen and his contemporaries used a metal scalpel** Bill Hayes, *Five Quarts: A Personal and Natural History of Blood* (New York: Random House, 2005): 172–73.

114 **In 1462 a bloodletting calendar** Starr, *Blood: An Epic History of Medicine and Commerce,* 19.

116 **Even drowning victims were bled** Wendy Moore, *The Knife Man* (New York: Broadway Books, 2005), 187–88.

5: The Red Stuff

130 **Hemoglobin is so effective at carrying O_2** Kenneth Walker, *The Story of Blood* (New York: Philosophical Library, 1962), 39.

131 **There are so many erythrocytes** Ibid., 37.

136 **A year later, encouraged by Lower's results** Bill Hayes, *Five Quarts: A Personal and Natural History of Blood* (New York: Random House, 2005), 52.

136–7 **Then he received about six ounces** Douglas Starr, *Blood: An Epic History of Medicine and Commerce* (New York: Knopf, 1998), 3–16.

141 **Aneurysms can occur for any** Robert and Michèle Root-Bernstein, *Honey, Mud, Maggots, and Other Medical Marvels* (New York: Houghton Mifflin, 1997), 78–79.

142 **Bloodletting was also used** Ibid., 80.

142 **In his book *Blood: An Epic History of Medicine and Commerce*** Ibid., 15–16.

143 **Porphyria (from the Greek word for "purple")** Matthew Bunson, *The Vampire Encyclopedia* (New York: Gramercy, 1993), 210.

144 **In the 1960s two authors** I. Macalpine and R. Hunter, "The Insanity of George III: A Classic Case of Porphyria," *British Medical Journal* 1 (1966): 65–67.

144 **The examination of several strands** BBC News, *King George III: Mad or Misunderstood,* http://news.bbc.co.uk/go/pr/fr/-/hi/health/388903.stm, 2004.

144 **Recently, researchers have found evidence** Andrew Bowser,

"DG Dispatch–DDW: Blood-letting Improves Hepatitis C Patient Response to Interferon," May 19, 1999, http://pslgroup.com/dg/fead6.htm.

145 **Studies have shown that insulin resistance** J. M. Fernandez-Real, G. Penarroja, A. Castro, F. Garcia-Bragado, I. Hernandez-Aguado, and W. Ricart, "Blood Letting in High-Ferritin Type 2 Diabetes: Effects on Insulin Sensitivity and Beta-Cell Function," *Diabetes* 51, no. 4 (2002): 1000–4.

6: A Beautiful Friendship

158 **One Trinidadian genus has** Roy T. Sawyer, *Leech Biology and Behavior*, vol. 1: *Anatomy, Physiology, and Behaviour* (Oxford, England: Clarendon Press, 1986), 368.

161 **Leeches are commonly fed upon** James H. Thorp and Alan P. Covich, eds., *Ecology and Classification of North American Freshwater Invertebrates* (New York: Academic Press, 1991), 428.

162 **Among these are several members** Roy T. Sawyer, *Leech Biology and Behavior*, vol. 2: *Feeding Biology, Ecology, and Systematics* (Oxford, England: Clarendon Press, 1986), 430–32.

162 **Although the hard outer covering** Thorp and Covich, *Ecology and Classification of North American Freshwater Invertebrates*, 451–52.

163 **Additionally, leeches were the preferred method** Robert and Michèle Root-Bernstein, *Honey, Mud, Maggots, and Other Medical Marvels* (New York: Houghton Mifflin, 1997), 90.

163 **Medicinal leech use reached its zenith** Ibid.

164 **Fashion-conscious ladies** *"Hirudo medicinalis,* Leech History, http://www.leeches-medicinalis.com/history.htm, 2006.

165 **The Leech-gatherers take them** J. G. Wood, *Animate Creation: A Popular Edition of Our Living World: a Natural History*, vol. 3 (New York: Selmar Hess, 1885), 598.

166 **Like many freshwater leeches** Sawyer, *Leech Biology and Behavior*, vol. 2: *Feeding Biology, Ecology, and Systematics*, 626–27.

168 **This behavior is similar to that reported** M. R. Heupel, C. A. Simpfendorfer, and R. E. Hueter, "Running Before the Storm: Blacktip Sharks Respond to Falling Barometric Pressure Associated with Tropical Storm Gabrielle," *Journal of Fish Biology* 63, no. 5 (2003): 1357–63.

170 **Leeches were also commonly used to treat strokes** A. Mark Clarfield, "Stalin's Death (or 'Death of a Tyrant')," *Annals of Long-Term Care* 13, no. 3 (March 2005): 52–54.

170 **Summoned, some contend, up to thirteen hours** Edvard Radzinsky, *Stalin* (New York: Doubleday, 1996), 574.

170 **Stalin had recently initiated** Ibid., 552–65.

171 **Presumably, this became a popular measure** A. Park, "The Case of the Disappearing Leech," *British Journal of Plastic Surgery* 46 (1993): 543.

172 **Leeches that specialize in larger hosts** Sawyer, *Leech Biology and Behavior*, vol. 2: *Feeding Biology, Ecology, and Systematics,* 454.

172 **Sucker attachment actually has two components** Sawyer, *Leech Biology and Behavior*, vol. 1: *Feeding Biology, Ecology, and Systematics,* 358.

174 **Similarly, other species of** H. E. Müller, M. Pinus, and Uwe Schmidt, "*Aeromonas hydrophila* as a Normal Intestinal Bacterium of the Vampire Bat, *Desmodus rotundus*," *Zentralblatt Für Veterinärmedzin Reihe B.* 27, no. 5 (1980), 419–24.

175 **In a 1994 study, Norwegians** Anders Baerheim and Hogne Sanvik, "Effect of Ale, Garlic, and Soured Cream on the Appetite of Leeches," *British Medical Journal* 309 (December 24, 1994): 1689.

177 **In the late nineteenth century** J. S. Haller Jr., "Decline of Bloodletting: A Study in 19th-Century Ratiocinations," *Southern Medical Journal* 79 (1986): 469–75.

7: Sleeping with the Enemy

192 ***Ecdysiast* was also the term coined** Joseph D. Ayd, "H. L., Where Are You? A Celebration of Henry Mencken on the Centennial of His Birth," *English Journal,* 69, no. 6 (1980): 32–37.

195 **Those of you looking** *Proceedings of the Royal Society of London B, Biological Sciences* 205 (1979): 581–98.

198 **Various insects and other arthropods** Robert L. Usinger, *Monograph of Cimicidae* (College Park, Md.: Entomological Society of America, 1966), 31–32.

199 **Finally, although an 1855 paper** Bruce Cummings, *The Bed-Bug: Its Habits and Life History and How to Deal with It,* 6th ed., Economic Series No. 5, British Museum (Natural History) (London: Adlard and Son, Limited, Bartholomew Press, Dorking) 1949, 17.

200 **Recently, scientist David Reed and his co-workers** David L. Reed, Jessica E. Light, Julie M. Allen, and Jeremy J. Kirchman, "Pair of Lice Lost or Parasites Regained: The Evolutionary History of Anthropoid Primate Lice," *BMC Biology* 5, no. 7 (March 7, 2007), doi:10.1186/1741-7007-5-7.

200 ***Monograph of Cimicidae*** Usinger, *Monograph of Cimicidae,* 1–7.

201 **Medicinal uses for bed bugs** Ibid., 7.

201 **Quintus Serenus was another Roman** Ibid.

201 **According to Usinger** Ibid.

202 ***A Treatise of Buggs*** John Southall, *A Treatise of Buggs* (London, 1730).

203 **Southall's interviews supported the claims** Ibid., 3.

203 **In this regard, the Yanks were** Cummings, *The Bed-Bug: Its Habits and Life History and How to Deal With It,* 3.

204 **Currently, scientists recognize around** Usinger, *Monograph of Cimicidae,* 1.

204 **Reflecting their worldwide distribution** Ibid, 4–5.

204 **Besides "red coats" and "heavy dragoons"** Cummings, *The Bed-Bug: Its Habits and Life History and How to Deal With It,* 3.

206 **Harkening back to the enormous** Ibid., 12.

209 **Speaking of bugs, the English word** Usinger, *Monograph of Cimicidae,* 5.

215 **Fortunately, in some states** New York State, Department of State, Division of Licensing Services, "Manufacture, Repairer-Renovator or Rebuilder of New and/or Used Bedding and/or Retailer/Wholesaler of Used Bedding Application," http://www.dos.state.ny.us/lcns/instructions/1427ins.html.

8: Of Mites and Men

233 **During World War II** Tyler A. Woolley, *Acarology: Mites and Human Welfare* (New York: John Wiley and Sons, 1988), 444.

234 **With no specific treatment available** D. J. Kelly, A. L. Richards, J. Temenak, D. Strickman, and G. A. Dasch. "The Past and Present Threat of Rickettsial Diseases to Military Medicine and International Public Health," *Clinical Infectious Disease* 34, Suppl. 4 (2002): S145–69.

234 **All along the Papuan coast** Emory C. Cushing, *History of Entomology in World War II* (Pub. 4294). (Washington, D.C.: Smithsonian Institution, 1957), 80–81.

234 **Eventually, antibiotics like tetracycline, doxycycline, and chloramphenicol** George Watt and David Walker. "Scrub Typhus," in *Tropical Infectious Diseases: Principles, Pathogens, and Practice,* vol. 1, ed. Richard Guerrant, David H. Walker, and Peter Weller, 592–97. (Philadelphia: Churchill Livingstone, 1999).

235 **In parts of northern Thailand** George Watt, C. Chouriyagune, R. Ruangweerayud, P. Watcharapichat, D. Phulsuksombati, K. Jongsakul, et al., "Scrub Typhus Infections Poorly Responsive to Antibiotics in Northern Thailand," *Lancet,* 348 (1996): 86–89.

237 **The basic premise, proposed by** Gavin de Beer, *Embryology and Evolution* (Oxford, England: Clarendon Press, 1930).

237 **and reinvigorated by Stephen** Stephen Jay Gould, *Ontogeny and Phylogeny* (Cambridge, Mass.: Belknap Press, 1977), 4.

243 **Among the acarids, perhaps the strangest** Timothy G. Myles, "Observations on Mites (Acari) Associated with the Eastern Subterranean Termites, *Reticulitermes flavipes* (Isoptera: Rhinotermitidae)," *Sociobiology* 39, no. 2 (2002): 277–80.

244 **In perhaps the strangest case of phoresy** R. K. Colwell, "Effects of Nectar Consumption by the Hummingbird Flower Mite *Proctolaelaps kirmsei* on Nectar Availability in *Hamelia patens,*" *Biotropica* 27 (1995): 206–17.

244 **Acarologist Tyler Woolley** Woolley, *Acarology: Mites and Human Welfare,* 3.

244 **According to entomologists R. Chapman and H. Shepard**

Arnold Mallis, *Handbook of Pest Control,* 2d ed. (New York: MacNair-Dorland Co., 1954), 863.

244 **And in a quote that immediately reminded** R. N. Chapman and H. H. Shepard, "Insects Infesting Stored Food Products," *University of Minnesota Agricultural Experiment Station Technical Bulletin* 198 (1932).

245 **For example, approximately 140** William Olkowski, Sheila Daar, and Helga Olkowski. *Common-Sense Pest Control* (Newtown, Ct.: Taunton Press, 1991), 159.

245 **Scabies is a condition that produces** Ibid., 164–66.

246 **Scabies is a disease of herding** John H. Stokes, "Scabies Among the Well-to-Do," *Journal of the American Medical Association* 106 (1936): 675.

246 ***Varroa* can be considered an invertebrate vampire** Gwilym O. Evans, *Principles of Acarology* (Wallingford, UK: CAB International, 1992), 173–74.

248 **In a pilot study published by the International Association** Wolfgang Harst, Jochen Kuhn, and Hermann Stever, "Can Electromagnetic Exposure Cause a Change in Behavior? *Acta Systemica–IIAS International Journal* 6, no. 1 (2005): 1–6.

249 **A number of researchers have** B. V. Ball and M. F. Allen, "The Prevalence of Pathogens in the Honeybee *(Apis mellifera)* Colonies Infected with the Parasitic Mite *Varroa jacobsoni, Annals of Applied Biology* 113 (1988): 337–44.

249 **These viruses are thought** J. R. de Miranda, M. Drebot, S. Tyler, M. Shen, C. E. Cameron, D. B. Stoltz, et al., "Complete Nucleotide Sequence of Kashmir Bee Virus and Comparison with Acute Bee Paralysis Virus, *Journal of General Virology* 85 (2004): 2263–70.

251 **Pellegrino gave another** "Bat Die-Off Prompts Investigation; DEC Asks for Cavers' Help to Prevent Spread of 'White Nose Syndrome,'" New York State Department of Environmental Conservation, http://www.dec.ny.gov/press/41621.html, January 30, 2008.

252 **Many of these species are relatively harmless** Evans, *Principles of Acarology,* 187–188.

254 **What saved civilization, apparently** Andrew B. Appleby,

"The Disappearance of the Plague: A Continuing Puzzle," *Economic History Review* 33, no. 2 (2004): 161–73.

256 **Hard-bodied ticks, ixodids** Evans, *Principles of Acarology,* 390.

260 **For example, the larval instars of** Ibid., 179.

264 **Similarly, Dr. Lauren Krupp and her colleagues** L. B. Krupp, L. G. Hyman, R. Grimson, P. K. Coyle, P. Melville, S. Ahnn, et al., "Study and Treatment of Post Lyme Disease (STOP-LD): A Randomized Double Masked Clinical Trial," *Neurology* 60 (2003): 1923–30.

264 **Alternatively, some researchers felt** A. C. Steere, E. Taylor, G. L. McHugh, and E. L. Logigian, "The Overdiagnosis of Lyme Disease," *Journal of the American Medical Association* 269, no. 14 (1993): 1812–16.

9: Candiru: Trouble with a Capital *C* and That Rhymes with *P*

271 **According to some of these** Stephen Spotte, *Candiru: Life and Legend of the Bloodsucking Catfish* (Berkeley, Calif.: Creative Arts Book Company, 2001), 157–66.

272 **These fishes are greatly attracted by the odor** Ibid., 157. From a translation in Carl H. Eigenmann, "The Pygidiidae, a Family of South American Catfishes," in *Memoirs of the Carnegie Museum* (Pittsburgh, Penn.: Carnegie Museum, 1918), 259–98.

272 **Fortunately, as with the candiru's fellow** Spotte, *Candiru: Life and Legend of the Bloodsucking Catfish.*

272 **Candiru belong to the Trichomycteridae** Warren Burgess, *An Atlas of Freshwater and Marine Catfishes* (Neptune City, N.J.: TFH Publications, 1993), 305–25.

272 **Trichomycterids, most of which are rather plain-looking** Spotte, *Candiru: Life and Legend of the Bloodsucking Catfish,* 5.

273 **Within the Trichomycteridae is a small subfamily** Ibid.

273 **According to Dr. Spotte** Ibid., 50–51.

273 **a technique detailed in an article published** Kenneth W. Vinton and W. H. Stickler, "The Carnero: A Fish Parasite of Man and Possibly Other Mammals," *Journal of Surgery* N.S. 54 (1941): 511–19.

275 **There have been numerous anecdotal descriptions** Paulo Petry, Anoar Samad, and Stephen Spotte, "Candiru Attack on Human in the Amazon River: Hard Evidence for a Long Standing Myth" (paper presented at the American Society of Herpetologists and Ichthyologists, July 6, 2001).

275 **A recent claim by researchers** Jansen Zuanon and Ivan Sazima, "Vampire Catfishes Seek the Aorta Not the Jugular: Candirus of the Genus *Vandellia* (Trichomycteridae) Feed on Major Gill Arteries of Host Fishes," *Journal of Ichthyology & Aquatic Biology* 8, no. 1 (2003): 31–36.

278 **In the "urine-loving hypothesis"** Spotte, *Candiru: Life and Legend of the Bloodsucking Catfish,* 142–49.

279 **Interestingly, these apparent drawbacks** Eigenmann, "The Pygidiidae, a family of South American Catfishes," 266–67.

279 **In what has become known** Spotte, *Candiru: Life and Legend of the Bloodsucking Catfish,* 154–56.

10: A Tough Way to Make a Living

283 **What was so thought provoking** Kurt Vonnegut, *Galápagos,* (New York: Delacorte Press, 1985).

286 **In that regard, they** Dolph Schluter and Peter Grant, "Ecological Correlates of Morphological Evolution in a Darwin's Finch, *Geospiza difficilis,*" *Evolution* 38, no. 4 (1984): 856–69.

286 ***Geospiza difficilis* is widely distributed** Peter Grant, B. Rosemary Grant, and Kenneth Petren, "The Allopatric Phase of Speciation: The Sharp-Beaked Ground Finch *(Geospiza difficilis)* on the Galápagos Islands," *British Journal of the Linnaean Society* 69 (2000): 287–317.

293 **We should judge every scrap of biodiversity** Edward O. Wilson, *The Diversity of Life* (Cambridge, Mass.: Belknap Press, 1992), 351.

SELECTED BIBLIOGRAPHY

Articles

Clarfield, A. Mark. "Stalin's Death (or 'Death of a Tyrant')." *Annals of Long-Term Care* 13, no. 3 (2005): 52–54.

Ditmars, Raymond L., and Arthur M. Greenhall. "The Vampire Bat—A Presentation of Undescribed Habits and Review of Its History." *Zoologica* 4 (1935): 53–76.

Goodwin, George, and Arthur M. Greenhall. "A Review of the Bats of Trinidad and Tobago," *Bulletin of the American Museum of Natural History*, 122 (1961): 187–301.

Gould, Steven J., and Richard Lewontin. "The Spandrels of San Marcos." *Proceedings of the Royal Society of London* B 205 (1979): 581–98.

Huxley, Thomas H. "On the Structure of the Stomach in Desmodus Rufus." *Proceedings of the Zoological Society of London* 35 (1865): 386–90.

Keegan, Hugh L., Myron G. Radke, and David A. Murphy. "Nasal Leech Infestation in Man." *American Journal of Tropical Medicine and Hygiene* 19, no. 6 (1970): 1029–30.

McFarland, William N., and William A. Wimsatt. "Renal Function and Its Relationship to the Ecology of the Vampire Bat, Desmodus Rotundus." *Comparative Biochemical Physiology* 28 (1970): 985–1006.

Mitchell, Clay G., and James R. Tigner. "The Route of Ingested Blood in the Vampire Bat," *Journal of Mammalogy* 51, no. 4 (1970): 814–17.

Myles, Timothy G., "Observations on Mites (Acari) Associated with the Eastern Subterranean Termites, *Reticulitermes flavipes* (Isoptera: Rhinotermitidae)." *Sociobiology* 39, no. 2 (2002): 277–80.

Park, A. "The Case of the Disappearing Leech." *British Journal of Plastic Surgery* 46 (1993): 543.

Schutt, William A., Jr., and J. Scott Altenbach. "A Sixth Digit in *Diphylla ecaudata,* the Hairy-Legged Vampire Bat." *Mammalia* 61, no. 2 (1997): 280–85.

Schutt, William A., Jr., John Hermanson, Young-Hui Chang, Dennis Cullinane, J. Scott Altenbach, Farouk Muradali, and John Bertram. "The Dynamics of Flight-Initiating Jumps in the Common Vampire Bat, *Desmodus rotundus.*" *Journal of Experimental Biology* 200, no. 23 (1997): 3003–12.

Schutt, William A., Jr., Farouk Muradali, Naim Mondol, Keith Joseph, and Kim Brockmann. "The Behavior and Maintenance of Captive White-Winged Vampire Bats, *Diaemus youngi* (Phyllostomidae: Desmodontinae)." *Journal of Mammalogy* 80, no. 1 (1999): 71–81.

Schutt, William A., Jr., and Nancy B. Simmons. "Morphology and Homology of the Chiropteran Calcar. *Journal of Mammalian Evolution* 5, no. 1 (1998): 1–32.

Steere, A. C., E. Taylor, G. L. McHugh, and E. L. Logigian, "The Overdiagnosis of Lyme Disease," *Journal of the American Medical Association* 269, no. 14 (1993): 1812–16.

Wilkinson, Gerald S. "Reciprocal Food Sharing in the Vampire Bat." *Nature* 308 (1984): 181.

Wimsatt, William A., and Anthony Geurriere. "Observations on the Feeding Capacities and Excretory Functions of Captive Vampire Bats." *Journal of Mammalogy* 43 (1962): 17–26.

Books

Altenbach, J. Scott. *Locomotor Morphology of the Vampire Bat, Desmodus rotundus.* Special Pub. No. 6, American Society of Mammalogists, 1979.

Brown, David E. *Vampiro—The Vampire Bat in Fact and Fantasy.* Silver City, N. Mex.: High-Lonesome Books, 1994.

Bunson, Matthew. *The Vampire Encyclopedia.* New York: Gramercy, 1993.

de Beer, Gavin. *Embryology and Evolution.* Oxford, England: Clarendon Press, 1930.

Cushing, Emory C. *History of Entomology in World War II.* Pub. No. 4294. Washington, DC: Smithsonian Institution, 1957.

Evans, Gwilym O. *Principles of Acarology.* Wallingford, UK: CAB International, 1992.

Florescu, Radu, and Raymond T. McNally. *Dracula—A Biography of Vlad the Impaler.* New York: Hawthorne Books, 1973.

Gould, Stephen Jay. *Ontogeny and Phylogeny.* Cambridge, Mass.: Belknap Press, 1977.

Gould, Stephen Jay. *Wonderful Life: The Burgess Shale and the Nature of History.* New York: W.W. Norton, 1989.

Greenhall, Arthur M., and Uwe Schmidt, eds. *Natural History of Vampire Bats.* Boca Raton, Fl.: CRC Press, 1988.

Hayes, Bill. *Five Quarts: A Personal and Natural History of Blood.* New York: Random House, 2005.

Kunz, Thomas H., and Paul Racey, eds. *Bat Biology and Conservation.* Washington, D.C.: Smithsonian Institute Press, 1998.

Moore, Wendy. *The Knife Man.* New York: Broadway Books, 2005.

Radzinsky, Edvard. *Stalin.* New York: Doubleday, 1996.

Reiss, Oscar. *Medicine and the American Revolution.* Jefferson, N.C.: McFarland and Co., 1998.

Root-Bernstei, Robert and Michèle. *Honey, Mud, Maggots, and Other Medical Marvels.* New York: Houghton Mifflin, 1997.

Sawyer, Roy T. *Leech Biology and Behavior,* vol. 1: *Anatomy, Physiology, and Behavior.* Oxford, England: Clarendon Press, 1986.

Sawyer, Roy T. *Leech Biology and Behavior,* vol. 2: *Feeding Biology, Ecology, and Systematics.* Oxford, England: Clarendon Press, 1986.

Sawyer, Roy T. *Leech Biology and Behavior,* vol. 3: *Bibliography.* Oxford, England: Clarendon Press, 1986.

Sigerist, Henry E. *A History of Medicine, vol. 1: Primitive and Archaic Medicine.* New York: Oxford University Press, 1951.

Southall, John. *A Treatise of Buggs.* London, 1730.

Spotte, Stephen. *Candiru: Life and Legend of the Bloodsucking Catfish.* Berkeley, Calif.: Creative Arts Book Company, 2001.

Summers, Montague. *The Vampire: His Kith and Kin.* London: Kegan Paul, Trench Trubner and Co., 1928.

Usinger, Robert L. *Monograph of Cimicidae.* College Park, Md.: Entomological Society of America, 1966.

Walker, Kenneth. *The Story of Blood.* New York: Philosophical Library, 1962.

Wilson, Edward O. *The Diversity of Life.* Cambridge, Mass.: Belknap Press, 1992.

Woolley, Tyler A. *Acarology–Mites and Human Welfare.* New York: John Wiley and Sons, 1988.

Newspaper and Magazine Articles

Altman, Mara. "Bed Bugs & Beyond." *Village Voice,* December 13–19, 2006.

Chan, Sewell. "Everything You Need to Know About Bedbugs but Were Afraid to Ask." *New York Times,* October 15, 2006.

Singer, Mark. "Night Visitors." *New Yorker,* April 4, 2004.

Internet Articles

BBC News. "King George III: Mad or Misunderstood," 2004, http://news.bbc.co.uk/go/pr/fr/-/hi/health/388903.stm.

"The Death of George Washington, 1799." EyeWitness to History, 2001, http://www.eyewitnesstohistory.com.

"George Washington: Eyewitness Account of His Death," 2003, http://www.doctorzebra.com/prez/z_x01death_lear_g.htm.

New York State, Department of State, Division of Licensing Services. "Manufacture, Repairer-Renovator or Rebuilder of New and/or Used Bedding and/or Retailer/Wholesaler of Used Bedding Application," http://www.dos.state.ny.us/lcns/instructions/1427ins.html.

ACKNOWLEDGMENTS

To my beautiful wife, Janet, and son, Billy—together you are the best thing that has ever happened to me. Thank you for your patience, love, and unwavering support.

Very special thanks to my literary agent, the wonderful and wise Elaine Markson. At the Elaine Markson Agency, I am indebted to Elaine's assistant, Gary Johnson, and I thank him for his advice and kindness.

At Harmony, I was incredibly lucky to have John A. Glusman edit my first book. I am also indebted to his superb team of coworkers, especially Anne Berry, Emily Lavelle, and Kira Walton.

To my dear friend and colleague, Patricia J. Wynne—my sincerest thanks and admiration, especially for your ability to bring my ideas to life with pen and ink and for urging me to "get another oar in the water."

I've been fortunate to have had several mentors in my educational and professional life. Most important, I am grateful to John W. Hermanson (Field of Zoology, Cornell University), who took a chance on me in 1990. As my graduate committee chairman, mentor, and friend, John not only taught me how to think like a scientist but also the value of figuring things out for yourself. At Cornell, I was guided by the talented tag team of John E. A. Bertram and Deedra McClearn—with a huge and welcome assist from James "Camuto Jim" Ryan (Hobart and William Smith College).

At my favorite place in the world, the American Museum of

Natural History (AMNH), bat biologist extraordinaire, Karl F. Koopman, was and will remain an inspiration to me and I am proud to have known him. Arthur M. Greenhall confirmed my initial hunch that "a vampire bat wasn't a vampire bat wasn't a vampire bat," and in doing so this funny little New Yorker set me on a career path that would eventually lead to this book. Nancy B. Simmons has been an unwavering supporter, superb collaborator, and trusted friend. It is *definitely* good to know the Queen. Also at the AMNH, my "younger brother" Darrin Lunde and I have spent many hours deep in discussion (scientific and otherwise), generally with our friends Ollie, Big Nick, and Aurora. Darrin was integral in encouraging me to write this book as well as correcting some initial boo-boos. Constant and welcome support is always provided by my colleagues and dear friends in the AMNH Mammalogy Department: Patricia Brunaur (goddess of all secretaries); Neil Duncan, Ross MacPhee, and Ruth O'Leary (for laughing at the jokes); Rob Voss; and Eileen Westwig. Very special thanks go to Mark Siddall and Louis Sorkin (Department of Invertebrates) for their time and for the wealth of information they provided me on leeches and bed bugs, respectively; and to Scott Schaefer (Ichthyology) for the candiru contacts. Thanks also to Mary DeJong (AMNH Library) for her help and kindness over the years, and to Mary Knight (for the support and for the great *Dracula* book).

In Trinidad, the incomparable Farouk Muradali was involved in every aspect of my bat research there. Farouk is truly an unsung hero in the field of bat biology, and I will always be grateful that for some reason he decided to take me under his wing. While acting as the head of the Forestry Division's Anti-Rabies Unit, Farouk and his crew (especially Amos Johnson, Keith Joseph, Naim Mondol, Partap Seenath, and Patrick Wallace) helped me capture bats (although once or twice they simply brought them to me in a two-liter Coke bottle). Not only were they incredibly generous with their time, but they also shared their trade secrets with me—all the while making sure that Janet and I felt at home in their wonderful country. Special thanks also to Mrs. Nadra Gian (head, Wildlife Section, Forestry Division, Ministry of Agriculture) and Mr. Kirk Amour (current head of Trinidad's Anti-Rabies Unit) for their patience and assistance. At the PAX Guest House (Mt. St. Benedict), my friends Gerard Ramsawak and his wife, Oda (and, recently, their daughter, Dominique), always make us feel like family whenever we stay with them.

My longtime friend, Charles Pellegrino, "learned me how to rite good." I hope to do the same for him one day.

My wonderful friend Leslie Nesbitt Sittlow spent many hours assisting me during library searches and other related research in New York City. She also spent weeks accumulating a section on blood recipes that appears on my Web site, darkbanquet.com.

At the Southampton College Summer Writer's Workshop, I am indebted to my literary mentor, the incredibly talented and equally wise Bharati Mukherjee. Special thanks also to the man most responsible for making that conference a success each year—Robert Reeves, for his encouragement, advice, friendship, and especially for teaching me so much about the craft of writing. Thanks also to Clark Blaise, Bruce Jay Friedman, and Frank McCourt and to my fellow students, especially the wildly talented Helen Simonson.

At the Cornell Cooperative Extension I thank Dr. Jody Gangloff-Kaufmann for the information on bed bugs and also for introducing me to Tamson Yeh, who was so helpful in providing me with ideas (and references) for my chapter on mites, chiggers, and ticks. I'm also grateful for her editorial comments on several early versions of this material.

At C. W. Post College, I am indebted to Matt Draud (Biology) and to Katherine Hill-Miller (Dean of Arts and Sciences), for her kindness and unwavering support. Additional thanks to my colleagues, especially Scott Carlin, Paul Forestell, Art Goldberg, Terry Jacob, Jeff Kane, and Howard Reisman.

The leech and bed bug chapters would have suffered greatly were it not for two equally memorable figures—Rudy Rosenberg (Leeches USA) and Andy Linares (Bug Off, New York City), respectively.

To Maria Armour, my student/undergrad assistant/graduate student/lab assistant/colleague and dear friend, my sincere thanks for all of your hard work and for *always* being there.

Special thanks to Alice Cooper and to Katherine Turman, producer of his radio show *Nights with Alice Cooper.*

Lastly, I'd like to thank the following individuals: Daniel Abram (Rancho Transylvania), Bob Adamo; Ricky Adams; J. Scott Altenbach; Natalie Angier; Domenic Anziano; Ted Arnold (Cornell Store); Cara Baker; Susan Barnard (Basically Bats); John Bodnar; Frank Bonaccorso; Mark Brigham; Donna Carpenter; Young-Hiu Chang; Dennis Cullinane; John de Cuevas; Rose DiMango; Angelo and Amelia DiDonato; Rose DiDonato; Betsy Dumont (University of Massachusetts–Amherst); Howard Evans (Cornell); Brock Fenton; Mo Fortes (Telegraph Office); Kim and Chris Grant (gorgeswebsites.com); Margaret and Tom Griffiths (NASBR); Graham Hawks; Roy Horst (NASBR); Rose Italiano; Kathy Kennedy; Tigga Kingston; Tom Kunz (Boston University); Rita Langdon; the Evil Leung Sisters (Mary and Mimi) and their wonderful mother, Siu Yung Leung; Vittorio Maestro; Carrie McKenna; Dawn Montalto; Stuart Parsons (Go All Blacks); the Peconic Land Trust (PeconicLandTrust.org); Harold and Florence Pedersen; Scott Pedersen; the Pellegrinoids (Ashley, Kyle, and Kelly); Paulo Petry; John Pierce; Dan Riskin; Jerry Ruotolo (my good friend and favorite photographer); Bobby and Dee Schutt; Chuck and Eileen Schutt; Herb Sherman; Edwin Spicka (my mentor at the State University of New York–Geneseo); Stephen Spotte; Mike and Carol Trezza (my other parents); Wilson Uieda; Janny van Beem; Mrs. D. Wachter—for listening to me patiently, thirty or so years ago, when I said I wanted to be a writer; and Carl Zimmer.

INDEX

M